SEMICONDUCTORS - GROWTH AND CHARACTERIZATION

Edited by **Rosalinda Inguanta**
and **Carmelo Sunseri**

Semiconductors - Growth and Characterization
http://dx.doi.org/10.5772/68094
Edited by Rosalinda Inguanta and Carmelo Sunseri

Contributors

Usha Philipose, Abhay Singh, Sreekanth Mandati, Bulusu V. Sarada, Suhash Ranjan Dey, Shrikant V. Joshi, Massimo Innocenti, Emanuele Salvietti, Andrea Giaccherini, Filippo Gambinossi, Maurizio Passaponti, Maria Luisa Foresti, Francesco Di Benedetto, S.R. Majid, Pei Yi Chan, Tatjana Gric, Aleksej Trofimov, Ortwin Hess, Victor Barsan, Denys Kurbatov, Oleksandr Dobrozhan, Petro Danilchenko, Anatoliy Opanasyuk, Giulia Massaglia, Marzia Quaglio

Published by InTech

Janeza Trdine 9, 51000 Rijeka, Croatia

Publishing Process Manager Marijana Francetic
Technical Editor SPi Global
Cover InTech Design team

First published March, 2018
Printed in Croatia
Legal deposit, Croatia: National and University Library in Zagreb

Additional hard copies can be obtained from orders@intechopen.com

Semiconductors - Growth and Characterization, Edited by Rosalinda Inguanta and Carmelo Sunseri
p. cm.
Print ISBN 978-953-51-3883-9
Online ISBN 978-953-51-3884-6

For EU product safety concerns:
IN TECH d.o.o., Prolaz Marije Krucifikse Kozulić 3, 51000 Rijeka, Croatia,
info@intechopen.com or visit our website at intechopen.com.

Meet the editors

Rosalinda Inguanta is currently an assistant professor of Applied Physical Chemistry at the University of Palermo. She received her PhD degree in 2008 at the University of Palermo. Her research interests include the synthesis of nanostructured electrochemical devices for electrochemical sensors, batteries, solar cells, and electrolysers and the synthesis of biocoatings to prevent the corrosion of medical devices in human body and the recovery of precious metals from electronic waste. She has developed a new method for the synthesis of nanostructures and thin film of different materials based on galvanic synthesis. In the field of semiconductors, she has studied the electrochemical deposition of different semiconductors, such as CIGS and CZTS; currently, her research team is employed on the study of perovskite thin film starting from electrodeposition of PbO_2 layers.

Carmelo Sunseri is a full professor of Applied Thermodynamic at the University of Palermo. His main research interests are in the fields of thermodynamic and kinetic study of electrochemical reactions on both metallic and semiconducting electrodes, electrochemical synthesis and characterization of nanostructured materials, and photoelectrochemical investigation of semiconductor surfaces. His research activity is focused on materials and systems of technological interest in view of their potential application in the following fields: electrochemical storage and conversion of energy (battery and fuel cells), photovoltaics, fabrication of thin-film devices, and electrochemical fabrication of porous devices.

Contents

Preface

The discovery of semiconducting materials allowed for tremendous and important advancements in the field of photonics, photovoltaics, electronics, and thermoelectrics. Semiconductors play a key role for the miniaturization of computers and manufacturing of computer parts. They are the beating heart of diodes, transistors, and photovoltaic cells.

Over the years, many efforts have been devoted to increase the knowledge on the numerous and fascinating properties of semiconducting materials, not only on silicon and germanium, certainly the most famous and used semiconductors, but also on multielement semiconductors such as chalcopyrite and kesterite. Furthermore, semiconducting plastics have attracted great attention in plastic light-emitting diodes (LEDs), which are flexible and can be molded to any desired shape.

This book deals with the synthesis and characterization of different semiconducting materials with special attention to the possible technological applications. Fundamental issues are also treated showing new developments about semiconductor physics.

Different deposition methods are discussed in this book, such as chemical bath deposition and spray pyrolysis for fabricating thin film and electrospinning process for obtaining semiconducting nanofibers. Particular attention has been devoted to the use of electrochemical methods for synthesizing semiconducting materials of interest for the hi-tech industry. To make semiconductor-based devices more attractive, low-cost and easily scalable processes must be developed. In this context, electrodeposition is a valuable tool because it is cheap, simple, and quick in comparison to other fabrication methods. Good-quality large-area thin films and nanostructures can be obtained by this technique. In particular, template-based electrochemical synthesis is of advantage for fabricating semiconductors with nanostructured morphology. A further advantage of the electrochemical route is the possibility to simultaneously control both electrical conduction type and energy gap of the semiconductors by adjusting the deposition parameters. In the case of multielement semiconductors, it is also possible to control the composition of deposit by adjusting the nature and composition of the deposition bath. Apart from conventional electrochemical methods such as direct electroplating and pulse or pulse-reverse electroplating, electrochemical atomic layer deposition (E-ALD) of metal chalcogenides is also discussed.

Rosalinda Inguanta and Carmelo Sunseri
Applied Physical Chemistry Laboratory
Department of Industrial and Digital Innovation –
Chemical, Computer, Management, Mechanical Engineering
University of Palermo
Palermo, Italy

Section 1

General Methods for Semiconductors Growth

Chapter 1

Investigation of the Nanostructured Semiconductor Metamaterials

Aleksej Trofimov, Tatjana Gric and Ortwin Hess

Additional information is available at the end of the chapter

http://dx.doi.org/10.5772/intechopen.72801

Abstract

The presence of electromagnetic waves on two-dimensional interfaces has been extensively studied over the last several decades. Surface plasmonic polariton (SPP), which normally exists at the interface between a noble metal and a dielectric, is treated as the most widely investigated surface wave. SPPs have promoted new applications in many fields such as microelectronics, photovoltaics, etc. Recently, it has been shown that by nanostructuring the metal surface, it is possible to modify the dispersion of SPPs in a prescribed manner. Herein, we demonstrate the existence of a new kind of surface wave between two anisotropic meta-materials. In contrast to extensively studied surface waves such as SPPs and Dyakonov waves, the surface waves supported by the nanostructured semiconductor metamaterial cross the light line, and a substantial portion at lower frequencies lies above the free-space light line. Consequently, the proposed structure will interact with the material via leaky waves.

Keywords: metamaterials, semiconductor, surface plasmon polaritons

1. Introduction

Plasmonics and the recent birth of metamaterials (MMs) [1–4] and transformation optics [5, 6] are currently opening a gateway to the development of a family of novel devices with unprecedented functionalities ranging from sub-wavelength plasmonic waveguides and optical nanoresonators [7] to superlenses, hyperlenses [8] and light concentrators [9]. Coupling between photons and surface plasmon polaritons (SPPs) [10, 11] is enabled by the periodically nanostructured metallic films allowing for exceptional and tunable optical properties determined by a combination of design geometry, the surrounding dielectric permittivity and the choice of metal [12, 13]. SPPs have promoted new applications in many fields such as microelectronics [14], photovoltaics [15], near-field sensing [16], laser technology [17, 18], photonics [19], meta-materials design [2], high-order harmonics generation [20] or charged particle acceleration [21]. Recently, it has been shown that by nanostructuring the metal surface, it is possible to modify the dispersion of SPPs or excite the SPPs in a prescribed manner [22, 23].

The process of replacement of the uniaxial medium by a biaxial crystal [24], an indefinite medium [25] and a structurally chiral material [26] may enforce the presence of hybrid surface waves with some parallel characteristics. In the latter case, a methodology developed by Tamm [27] was adopted seeking to find a new type of surface wave, called as Dyakonov-Tamm wave, as it combines the features of Dyakonov surface waves (DSWs) and Tamm states. The use of structured materials with extreme anisotropy provided a fertile background aiming to increase the range of directions of DSWs substantially, as it is compared with the rather narrow range observed with natural birefringent materials [28]. Especially, outstanding results take place if the metallic nanoelements are employed to the anisotropic structures, as it occurs, for example, with a simple metal-dielectric multilayer, a case where the angular range may surpass half of a right angle [29]. The propagation length of these DSWs is drastically limited by the penetration depth inside the lossy MM [30] as it is caused by the specific damping capacity of metals.

The examination of two different interfaces, i.e. MM/dielectric and MM/TCO, is of the particular importance. Surface waves of different kinds, including DSWs along with traditional-like SPPs, are examined. Contrarily, the introduction of MM/TCO interface leads to a transformation of the traditional-like SPPs. As a consequence, the new types of surface waves are found.

Moreover, hyperbolic metamaterials, being special kind of anisotropic metamaterial with dielectric tenor elements having the mixed signs, have attracted growing attention due to their ability to support very large wave vectors. Their exotic features give rise to many intriguing applications, such as sub-wavelength imaging [31, 32] and hyper-lens [33, 34] that are infeasible with natural materials. In this paper, we demonstrate the existence of a new kind of surface wave between two anisotropic metamaterials. In contrast to extensively studied surface waves such as SPPs and Dyakonov waves, whose in-plane wave vector is greater than that of the bulk modes, the surface waves supported by the nanostructured semiconductor metamaterial cross the light line, and a substantial portion at lower frequencies lies above the free-space light line, which typically separates non-radiative (bound) and radiative (leaky) regions.

2. Transparent conducting oxide (TCO)—dielectric composite heterostructure-based multilayer metamaterial

The structure of the metamaterial (MM) is shown in **Figure 1**, where d_{TCO} and d_d represent the thicknesses of TCO and dielectric layers, respectively. All the involved media are nonmagnetic, so the magnetic permeability of every medium is the same as that of vacuum. In our numerical calculation, we use a semi-infinite TCO/PbS MM as an example to explore the dispersive features of hyperbolic surface plasmon polaritons (HSPPs), where ε_d = 18.8 for PbS layers, and ε_{TCO} is calculated using the parameters presented in [35].

http://dx.doi.org/10.5772/intechopen.72801

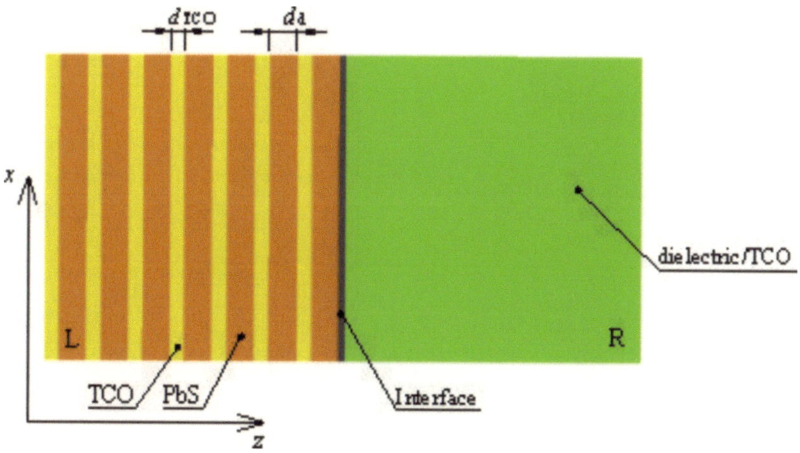

Figure 1. One-dimensional MM composed of alternating TCO and dielectric layers. The MM/dielectric ($\varepsilon_R = 1$ or $\varepsilon_R = 2.25$) and MM/TCO interfaces are considered.

Consequently, an example to estimate the limitation on the structure period under the effective-medium theory (EMT) is considered. Proposing model, when the wavelength of radiation is much larger than the thickness of any layer, one can apply the effective-medium approach based on averaging the structure parameters.

First, the dispersion features of HSPPs are investigated. On the contrary to the approach presented in [36–38], the damping term in the TCO is not ignored in the process of analysing and calculating their dispersion properties. It is worth to mention that this particular MM is equivalent to a uniaxial-anisotropy effective medium, with its anisotropy axis (the optical axis) along the structure periodicity in the long-wavelength limit. Its effective permittivity tensor is written as

$$\varepsilon = \varepsilon_0 \begin{pmatrix} \varepsilon_\parallel & 0 & 0 \\ 0 & \varepsilon_\parallel & 0 \\ 0 & 0 & \varepsilon_\perp \end{pmatrix}, \tag{1}$$

in the principal-axis coordinate system. The principal values of the tensor are expressed with [39, 40]

$$\varepsilon_\parallel = f_{TCO}\varepsilon_{TCO} + f_d\varepsilon_d \tag{2}$$

$$\varepsilon_\perp = \varepsilon_{TCO}\varepsilon_d/\left(f_{TCO}\varepsilon + f_d\varepsilon_{TCO}\right), \tag{3}$$

where $f_{TCO} = d_{TCO}/L$ and $f_d = d_d/L$, $L = d_d + d_{TCO}$ represent the TCO and dielectric filling ratios, respectively. It should be realized that MMs with a very large permittivity or a near-zero permittivity exhibit interesting properties [40]. While one may consider MMs with a very large permittivity as optical conductors, those with a near-zero permittivity can be used as optical insulators [40]. The zero point and divergence point of the principal values were also of

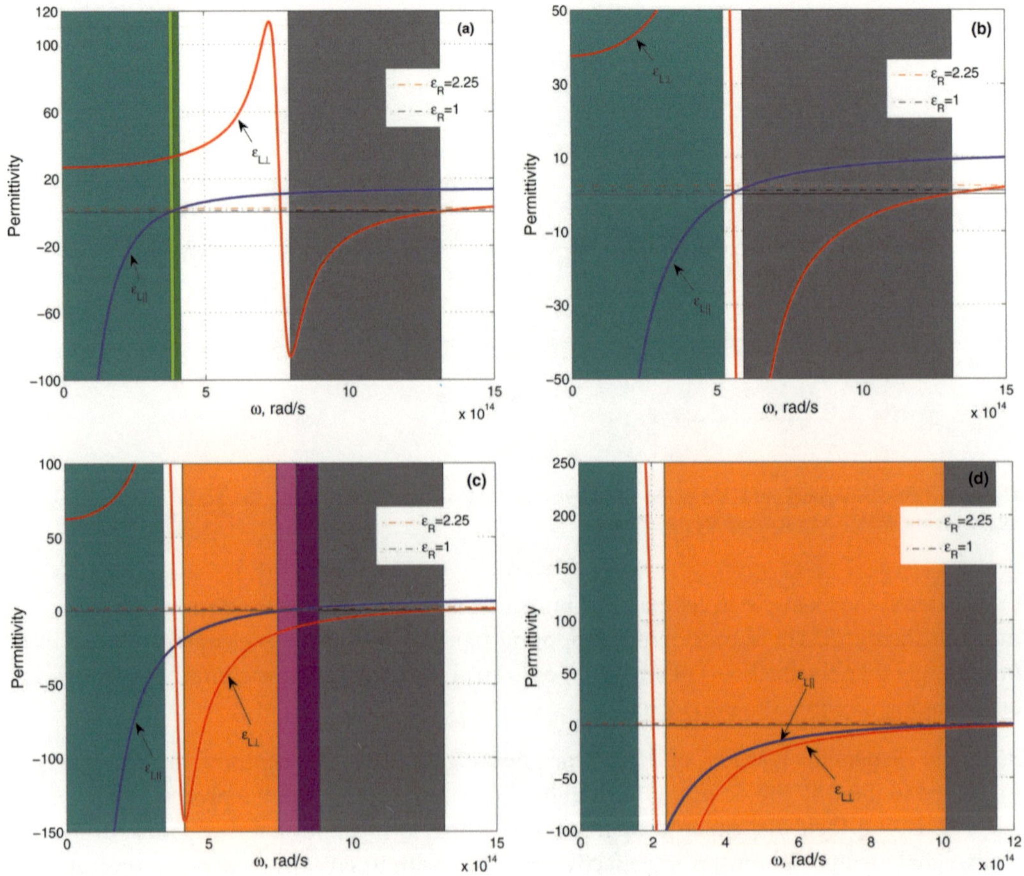

Figure 2. The principal values of effective permittivity and the different frequency regions for typical TCO-filling ratios corresponding to the MM/dielectric interface. $f_{TCO} = 0.3$ in (a), $f_{TCO} = 0.5$ in (b), $f_{TCO} = 0.7$ in (c) and $f_{TCO} = 0.9$ in (d).

the particular interest to discuss the dispersion features of SPPs [41]. The zero point and divergence point of $\varepsilon_{\perp}$ or $\varepsilon_{\|}$ will be applied to discuss HSPPs.

Electric and magnetic fields' tangential components need to be evaluated at the interface in order to get metamaterial interface confined surface mode unique dispersion [42]:

$$\beta = k\sqrt{\frac{\left(\varepsilon_{\|}^{R} - \varepsilon_{\|}^{L}\right)\varepsilon_{\perp}^{R}\varepsilon_{\perp}^{L}}{\varepsilon_{\perp}^{R}\varepsilon_{\|}^{R} - \varepsilon_{\perp}^{L}\varepsilon_{\|}^{L}}}, \tag{4}$$

where k is the absolute value of wave vector in vacuum and β is the component of the wave vector parallel to the interface.

It is interesting to notice that in the case of the MM interface, the obtained result for the dispersion is as follows:

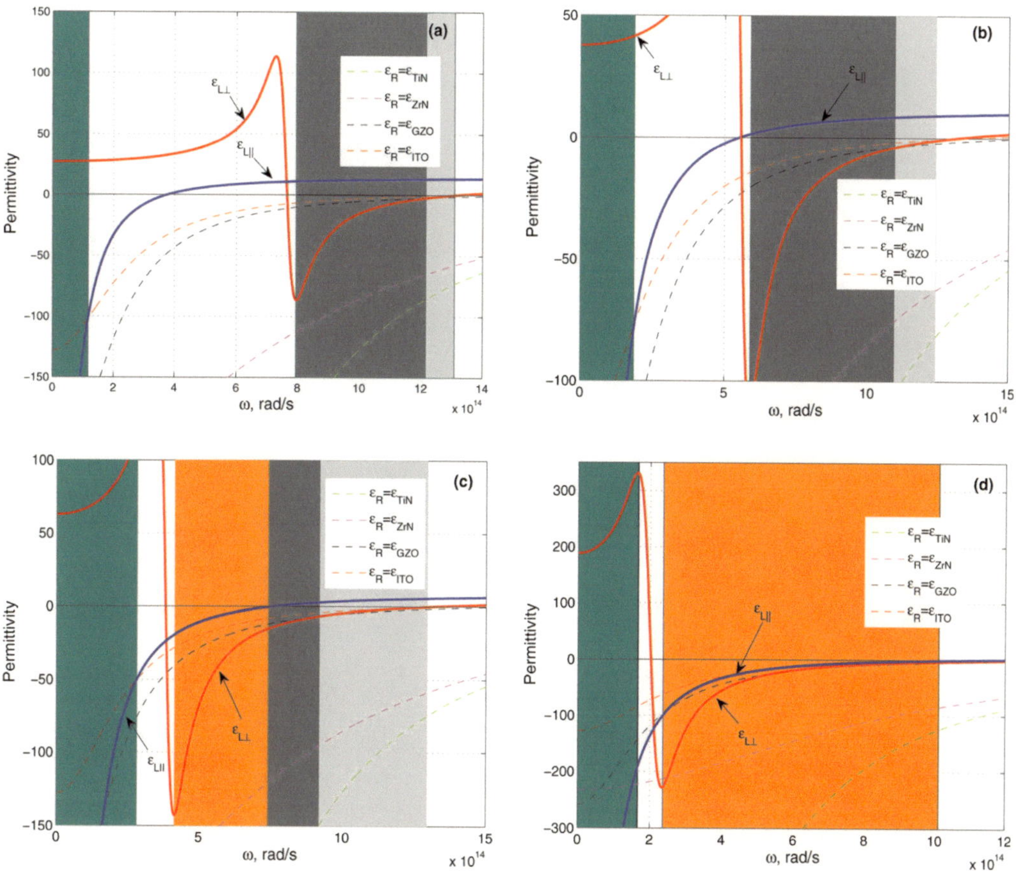

Figure 3. The principal values of effective permittivity and the different frequency regions for typical TCO-filling ratios, corresponding to the MM/TCO interface. $f_{TCO} = 0.3$ in (a), $f_{TCO} = 0.5$ in (b), $f_{TCO} = 0.7$ in (c) and $f_{TCO} = 0.9$ in (d).

$$\beta = k\left(\frac{\varepsilon_{TCO}\varepsilon_d\varepsilon_R\left(d_d - \frac{d_d f_{TCO}}{f_{TCO}-1}\right)\left(\varepsilon_{dR} - \frac{a}{d_d - \frac{d_d f_{TCO}}{f_{TCO}-1}}\right)}{\left(\varepsilon_R^2 - \frac{\varepsilon_{TCO}\varepsilon_d a}{b}\right)b}\right)^{1/2}, \qquad (5)$$

where $a = d_d\varepsilon_d - \frac{d_d\varepsilon_{TCO}f_{TCO}}{f_{TCO}-1}$, $b = d_d\varepsilon_{TCO} - \frac{d_d\varepsilon_d f_{TCO}}{f_{TCO}-1}$, ε_R is the permittivity of the material at the right-hand side of the interface.

It is of particular interest to obtain the dispersion relation for the interface states in the effective media approach corresponding to the MM interface, having in mind that material at the right-hand side is the same as employed in the MM. This dispersion relation reads

$$\beta = k\left(\frac{\varepsilon_{TCO}\varepsilon_d}{\varepsilon_{TCO} + \varepsilon_d}\right)^{1/2} \qquad (6)$$

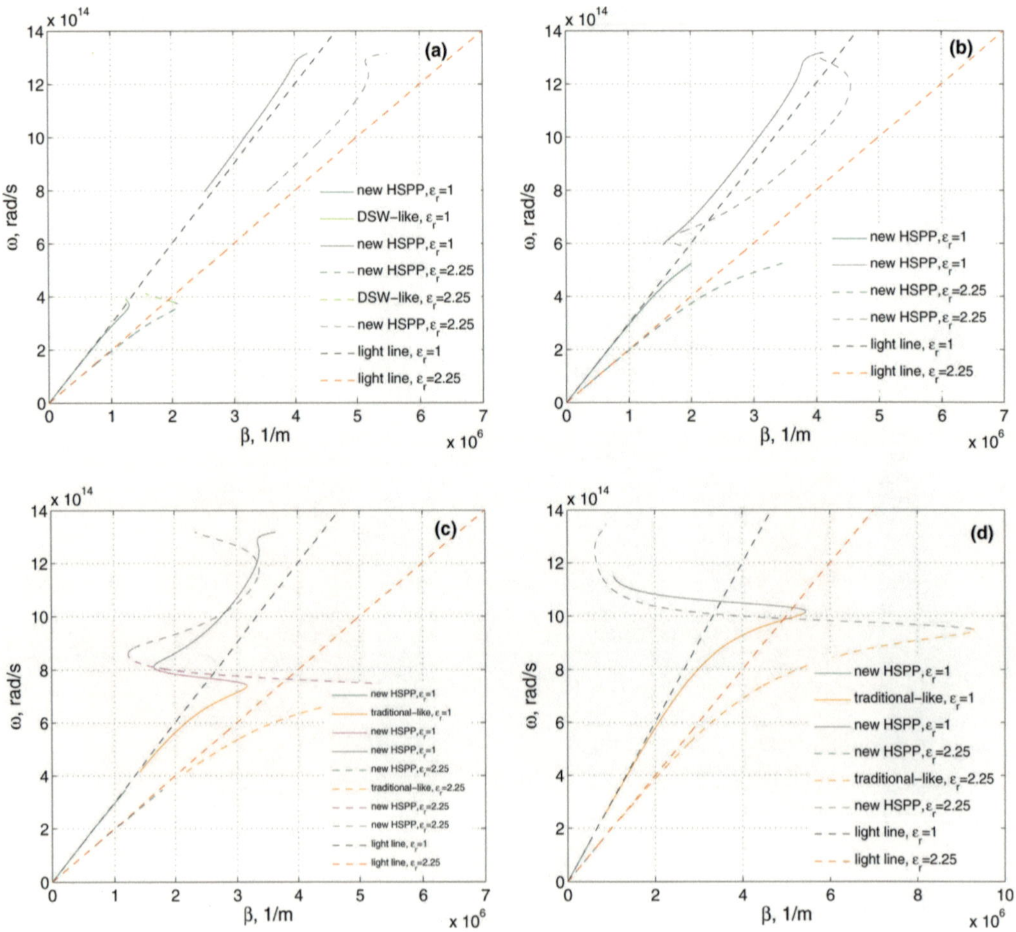

Figure 4. Dispersion curves of HSPPs for various TCO-filling ratios, i.e. $f_{TCO} = 0.3$ in (a), $f_{TCO} = 0.5$ in (b), $f_{TCO} = 0.7$ in (c) and $f_{TCO} = 0.9$ in (d), corresponding to the MM/dielectric interface.

As a matter of fact, we obtain a surprising result: the dispersion of a (single) interface mode does not depend on the thicknesses of the layers, and it coincides with the dispersion of a conventional surface plasmon at metal-dielectric interface.

2.1. The mode structure

In the case of a spatially infinite anisotropic material, invariant in two directions, the electromagnetic wave dispersion can be plotted for both MM/dielectric and MM/TCO cases. Thus, herein, we present analysis performed after the homogenization of the MM corresponding to the MM/dielectric and MM/TCO interfaces. Doing so, in the numerical calculations, the semi-infinite AZO/PbS MM is taken as an example. We will first review the optical properties by depicting the curves of the principal values ($\varepsilon_{\perp}$ and $\varepsilon_{\parallel}$) and the dielectric constant (ε_R) and then by distinguishing different frequency regions according to the properties of and relation

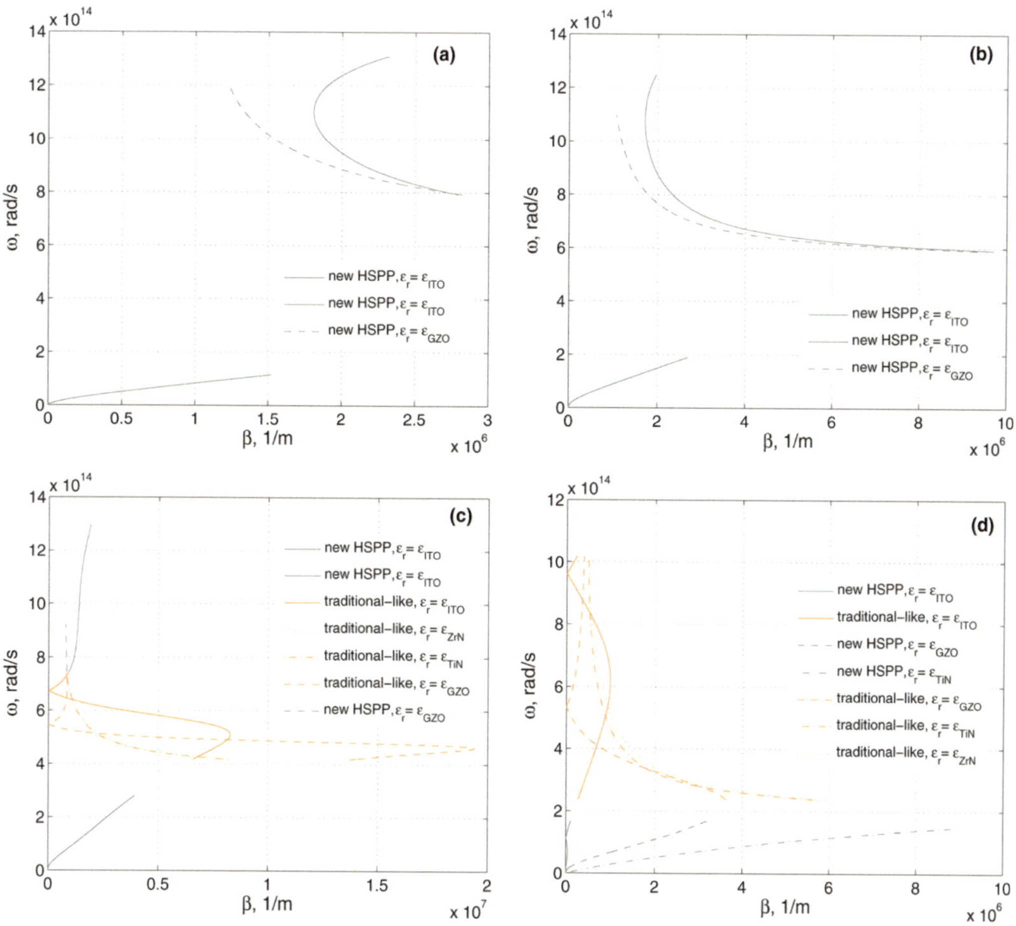

Figure 5. Dispersion curves of HSPPs for various TCO-filling ratios, i.e. $f_{TCO} = 0.3$ in (a), $f_{TCO} = 0.5$ in (b), $f_{TCO} = 0.7$ in (c) and $f_{TCO} = 0.9$ in (d), corresponding to the MM/dielectric interface.

among $\varepsilon_\perp$, $\varepsilon_\parallel$ and ε_R as shown in **Figures 2** and **3**. $\varepsilon_\perp > \varepsilon_R > \varepsilon_\parallel$ and $\varepsilon_\parallel < 0$ in the cyan region, $\varepsilon_\perp > \varepsilon_R > \varepsilon_\parallel$ in the green region, $\varepsilon_\parallel > \varepsilon_R > \varepsilon_\perp$ and $\varepsilon_\perp < 0$ in the grey region, $\varepsilon_R > \varepsilon_\parallel > \varepsilon_\perp$ and $\varepsilon_\perp < 0$ in the magenta region, and $\varepsilon_\perp < \varepsilon_\parallel < 0$ in the orange region. The tunability of the effective parameters presented in **Figures 2** and **3** has an effect on the dispersion curves of the HSPPs. As shown in **Figures 2** and **3**, the parameter that principally defines the tunability of the effective optical parameters of our metamaterial is the TCO-filling ratio, f_{TCO}. The dispersion curves are illustrated in **Figures 4** and **5** for various TCO-filling ratios.

The curve colors correspond to different frequency regions. For f_{TCO} = 0.3, there are three HSPPs belonging to three different kinds, respectively, in the case of MM/dielectric interface and two kinds of HSPPs in the case of MM/TCO interface. The upper waves exist in the gray region for MM/TCO case. Three exceptional cases related to f_{TCO} = 0.5, 0.7 and 0.9. In the case of f_{TCO} = 0.5, two kinds of HSPPs exist in two color regions, i.e. cyan and grey for MM/dielectric and MM/TCO interfaces. The upper short curves in the case of f_{TCO} = 0.7 lie in the gray region,

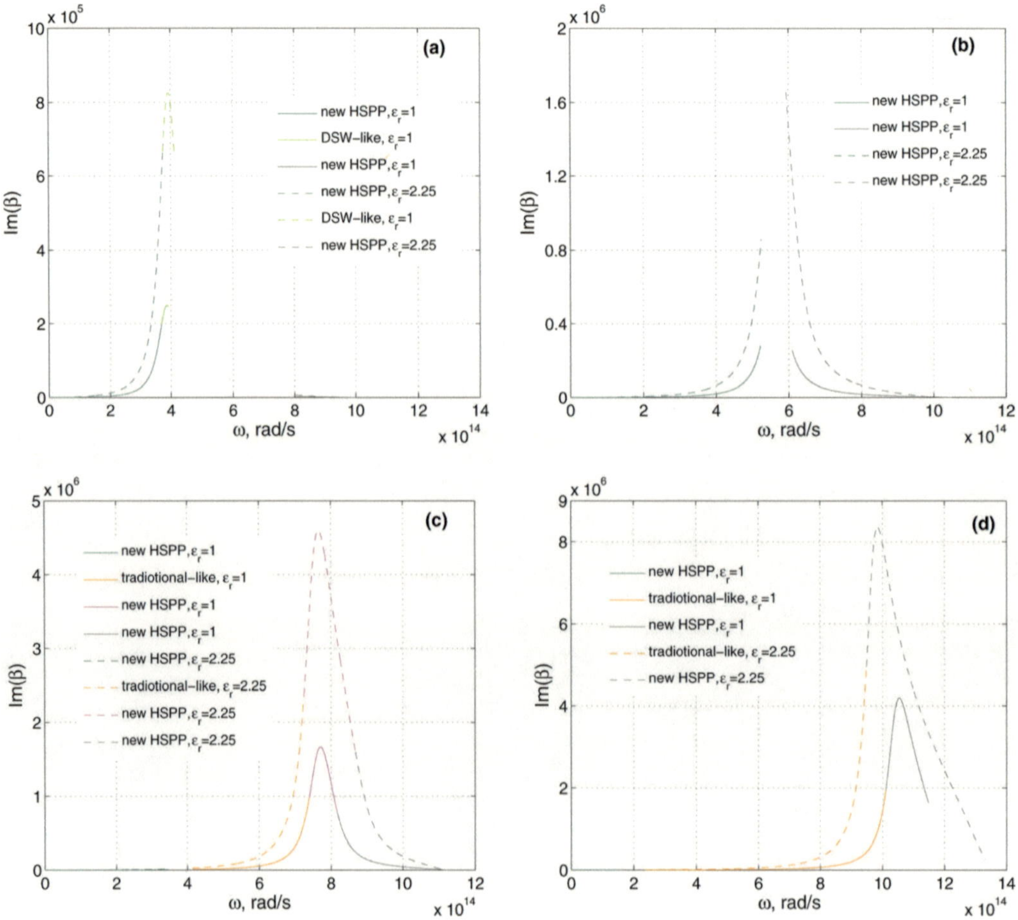

Figure 6. Absorption curves of HSPPs for various TCO-filling ratios, i.e. $f_{TCO} = 0.3$ in (a), $f_{TCO} = 0.5$ in (b), $f_{TCO} = 0.7$ in (c) and $f_{TCO} = 0.9$ in (d), corresponding to the MM/dielectric interface.

as illustrated in **Figures 4** and **5**. The other HSPPs lie in the orange region. The additional HSSPs lying in the magenta region correspond to the case of MM/dielectric interface. The extension of the magenta range for the case $\varepsilon_R = 2.25$ is displayed by the dark-magenta color. In the case of $f_{TCO} = 0.9$, two kinds of HSPPs are found lying in the orange and cyan regions for MM/TCO case and three kinds of HSPPS lying in the cyan, orange and gray regions for MM/dielectric case. **Figure 4** also demonstrates that there always is one HSPP in the cyan region for various TCO-filling ratios in MM/dielectric case. It is worthwhile mentioning that the case $\varepsilon_R = \varepsilon_{ITO}$ also allows for the rich phenomenon as the HSPP always exists in the cyan region for various filling ratios (**Figure 5**).

Based on the necessary condition for the existence of the DSW [39], the HSPP in the green region (**Figure 4**) is similar to the DSW so that it should be called Dyakonov-like SPP [2] or the Dyakonov defined in [39]. Moreover, the frequency range of the DSW existence can be extended by replacing the material at the right-hand side of the interface with $\varepsilon_R = 2.25$

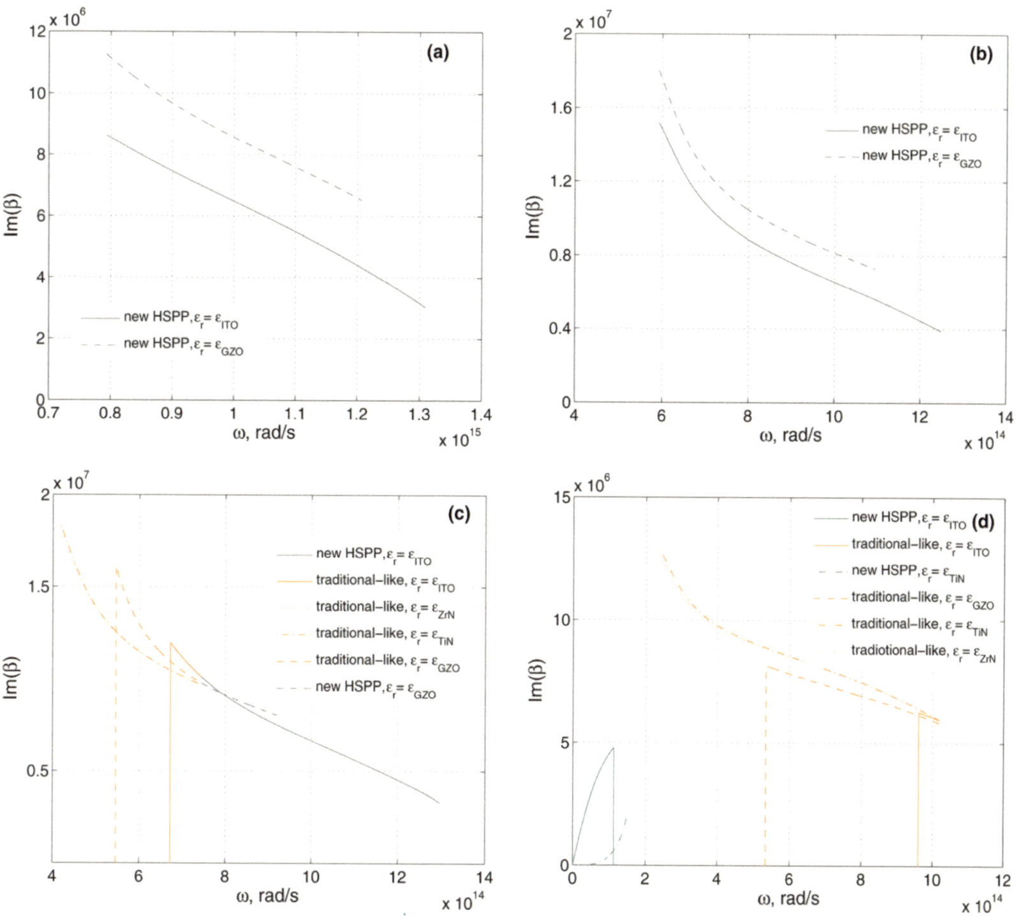

Figure 7. Absorption curves of HSPPs for various TCO-filling ratios, i.e. $f_{TCO} = 0.3$ in (a), $f_{TCO} = 0.5$ in (b), $f_{TCO} = 0.7$ in (c) and $f_{TCO} = 0.9$ in (d), corresponding to the MM/TCO interface.

(**Figures 2** and **4**). The former extension is demonstrated by the dark-green color in **Figure 2**. It is worthwhile mentioning that extension of the gray region in **Figure 3** is possible due to the employment of ITO instead of GZO at the right-hand side of the interface (the extension of the gray range is depicted by the light-gray color). Due to the negative principal values of the effective permittivity in the orange region, the HSPP in this region is similar to the traditional SPP, so we should name it the traditional-like SPP. The others are new types of HSPPs. Thus, all the HSPPs are divided into five kinds, situated in the five color regions in **Figures 4** and **5**, respectively. In the case of AZO/PbS MM and air/dielectric interface, five types of the HSPPs exist, and only three are seen in the case of AZO/PbS MM and TCO interface.

As expected, the traditional-like SPP has a typical dispersion in the case of the MM/dielectric interface, lying to the right of the light line. At the same time, the degradation of the dispersion in the orange region takes place in the case of the MM/TCO interface. The dispersion properties can be tuned with the TCO-filling ratio of the MM realization. While for low TCO-filling ratio in the case of the MM/TCO interface, two types of the modes are always present, for higher TCO-filling ratio, the disappearance of modes in the gray region takes place (**Figure 5(d)**).

The complex mode structure (**Figures 4** and **5**) corresponding to either MM/dielectric or MM/TCO interface emerges as a consequence of the confinement of plasmon polaritons in the direction perpendicular to the wave propagation. These electromagnetic surface waves arise via the coupling of the electromagnetic fields to oscillations of the conductor's electron plasma.

The imaginary parts of the wave vector (i.e. absorption) are plotted in **Figures 6** and **7**. It should be mentioned that negative values of the absorption in **Figures 6** and **7** result from non-physical solutions of the dispersion equation and have been omitted in line with [43]. Taking advantage of the absorption resonances, one can show that the simple multilayer structures without possessing any periodic corrugation have the prospective to act as directive and monochromatic thermal sources [44].

3. Nanostructured semiconductor metamaterial

Another interesting MM structure depicted in **Figure 8** is periodic stack of semiconductor-dielectric layers called hyperbolic metamaterial heterostructure.

The effective permittivity of the semiconductor (Si) can be calculated as follows:

$$\varepsilon_{1,3}(\omega) = \varepsilon_\infty - \frac{\omega_p^2}{\omega^2 + i\delta\omega}, \tag{7}$$

where ε_∞ is the background permittivity and ω_p is the plasma frequency. The effective-medium approach [45] which is justified if the wavelength of the radiation considered is much larger

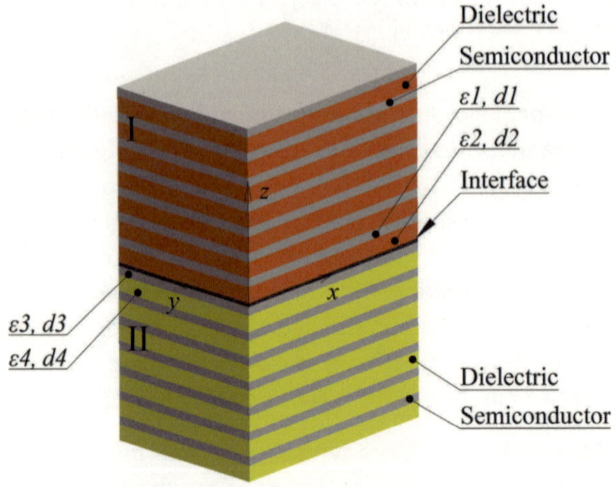

Figure 8. Geometry of the HMM. An interface separating two different semiconductor-dielectric-layered structures. Herein, indexes "1 and 2" correspond to the semiconductor and dielectric layers correspondingly.

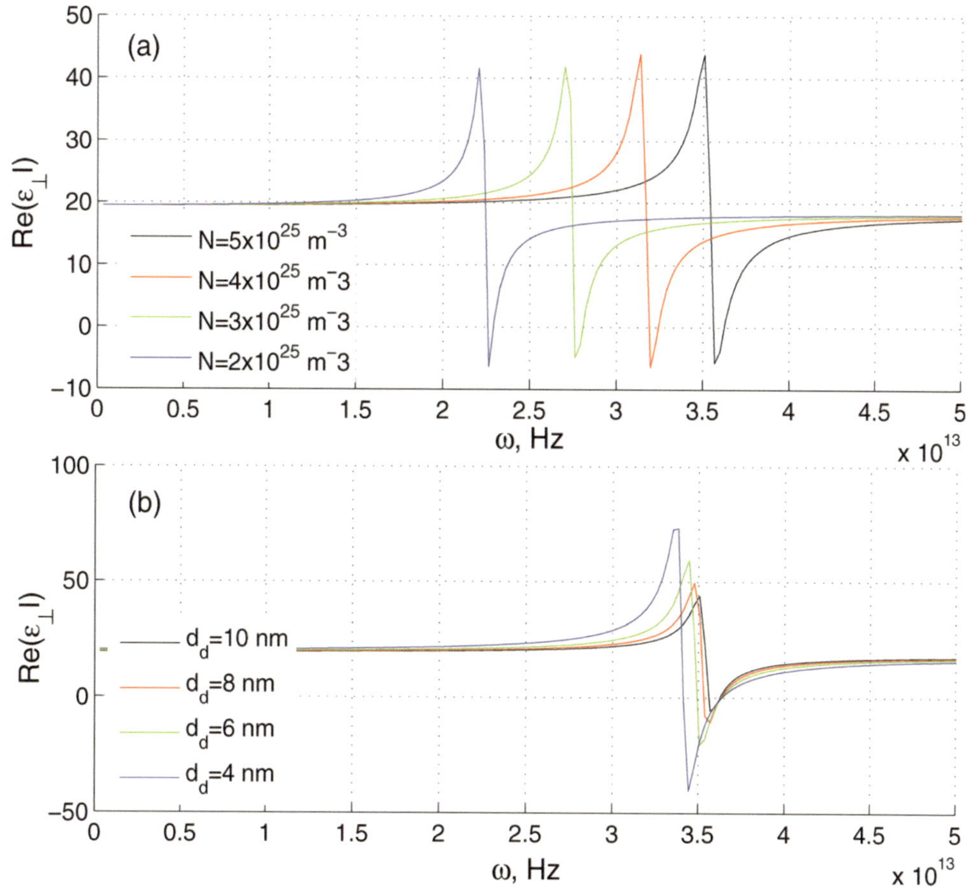

Figure 9. The influence of (a) doping level N and (b) thickness of dielectric d_d, on the real part of $\varepsilon_\perp$. $d_d = 10$ nm in (a) and $N = 5 \times 10^{25}$ m^{-3} in (b).

than the thickness of any layer is applied aiming to describe the optical response of such a system. The dispersion relation for the surface modes localized at the boundary separating two anisotropic media [42] is found by applying the appropriate boundary conditions, i.e. matching the tangential components of the electrical and magnetic fields at the interface. It is interesting to note that heavily doped silicon ($n > 2.2 \times 10^{19}$ cm^{-3}) has been shown to exhibit metallic properties at terahertz frequencies [46, 47] and has the potential to replace metals in such applications [48]. The case of a heavy-doped Si is considered, assuming that the doping level is $N = 5 \times 10^{19}$ cm^{-3} [49].

It should be mentioned that dramatic control of the frequency range of the surface wave existence is mostly concerned with the modifications of the permittivities and thicknesses of the layers [50] employed in the HMMs. To further study the surface waves, the tangential components of the electric and magnetic fields at the interface should be evaluated, and a single surface mode with the propagation constant should be obtained aiming to get the unique dispersion relation for the surface modes confined at the interface between two metamaterials [42].

Using the (4) formula, we can describe the case $\varepsilon_1 = \varepsilon_3$, $\varepsilon_2 = \varepsilon_4$, $d_1 \neq d_2 \neq d_3 \neq d_4$ reveals the dispersion as follows:

$$\beta = k\sqrt{\frac{\varepsilon_1\varepsilon_2}{\varepsilon_1 + \varepsilon_2}}. \tag{8}$$

The dispersion for the case of $\varepsilon_1 = \varepsilon_3$ and $\varepsilon_2 \neq \varepsilon_4$, $d_1 \neq d_2 \neq d_3 \neq d_4$ is as follows:

$$\beta = k\sqrt{\frac{\varepsilon_1^2\varepsilon_2\varepsilon_4(d_1+d_2)(d_3+d_4)\left(\frac{D}{d_1+d_2} - \frac{B}{d_3+d_4}\right)}{\left(\frac{\varepsilon_1\varepsilon_2 D}{C} - \frac{\varepsilon_1\varepsilon_4 B}{A}\right)CA}}, \tag{9}$$

where $A = d_4\varepsilon_1 + d_3\varepsilon_4$, $B = d_3\varepsilon_1 + d_4\varepsilon_4$, $C = d_1\varepsilon_2 + d_2\varepsilon_1$ and $D = d_1\varepsilon_1 + d_2\varepsilon_2$.

In the case of $\varepsilon_1 = \varepsilon_3$ and $\varepsilon_2 \neq \varepsilon_4$, $d_1 = d_3$ and $d_4 = d_2$:

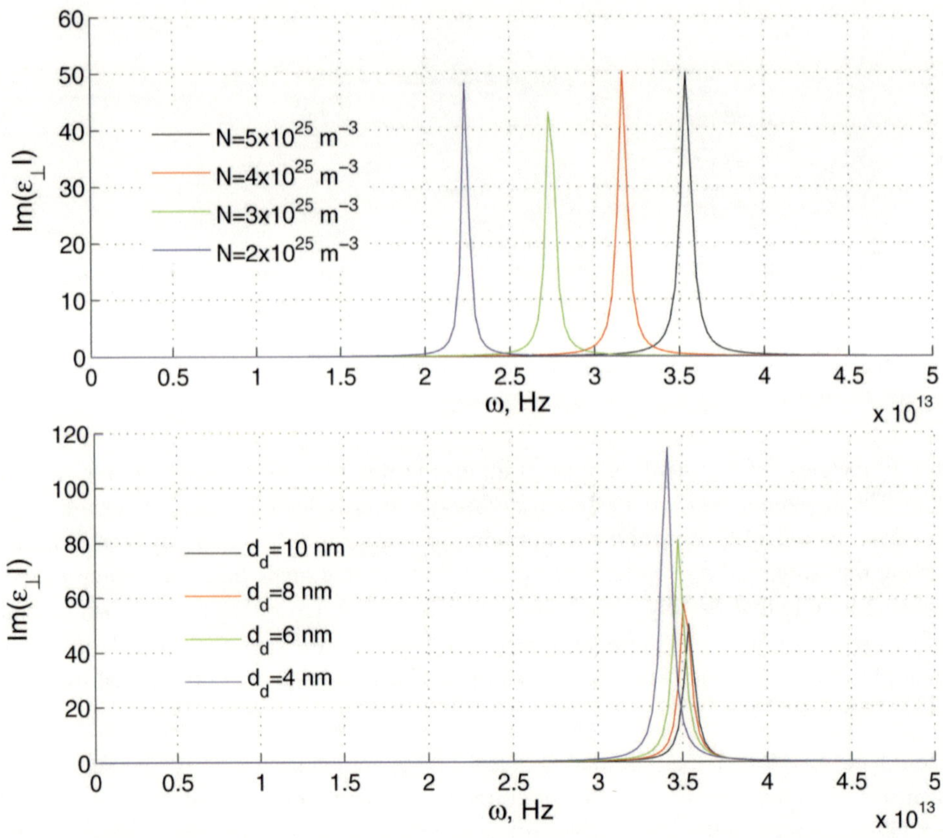

Figure 10. The influence of (a) doping level N and (b) thickness of dielectric d_d, on the imaginary part of $\varepsilon_\perp$. $d_d = 10$ nm in (a) and $N = 5 \times 10^{25}$ m^{-3} in (b).

$$\beta = k\sqrt{\frac{\varepsilon_1\varepsilon_2\varepsilon_4(d_1+d_2)}{d_1\varepsilon_1^2+d_2\varepsilon_1\varepsilon_2+d_1\varepsilon_2\varepsilon_4+d_2\varepsilon_1\varepsilon_4}} \tag{10}$$

The permittivity spectra for the perpendicular components of the considered multilayer heterostructure are demonstrated in **Figures 2(c), 3(a)–(c)** and **9(a)**. Tuning the doping level of the semiconductor may open a gateway to the frequency control of the hyperbolic dispersion curve as shown in **Figures 3(a), 4(a), 5(a)** and **9(a)**. It is assumed that d_s = 0.35 nm. PbS with relative permittivity ε_d = 18.8 and slab thickness t_d = 10 nm is chosen as the dielectric layer. It is clear that one has the potential to achieve the resonant behaviour of $\varepsilon_\perp$ by varying the doping level; moreover, the increase in the doping level causes a tuning of the resonant frequencies over the higher frequency range. Because of these attractive properties, our semiconductor-based layered structure has the great potential in the application as the building block for various HM-based optical devices.

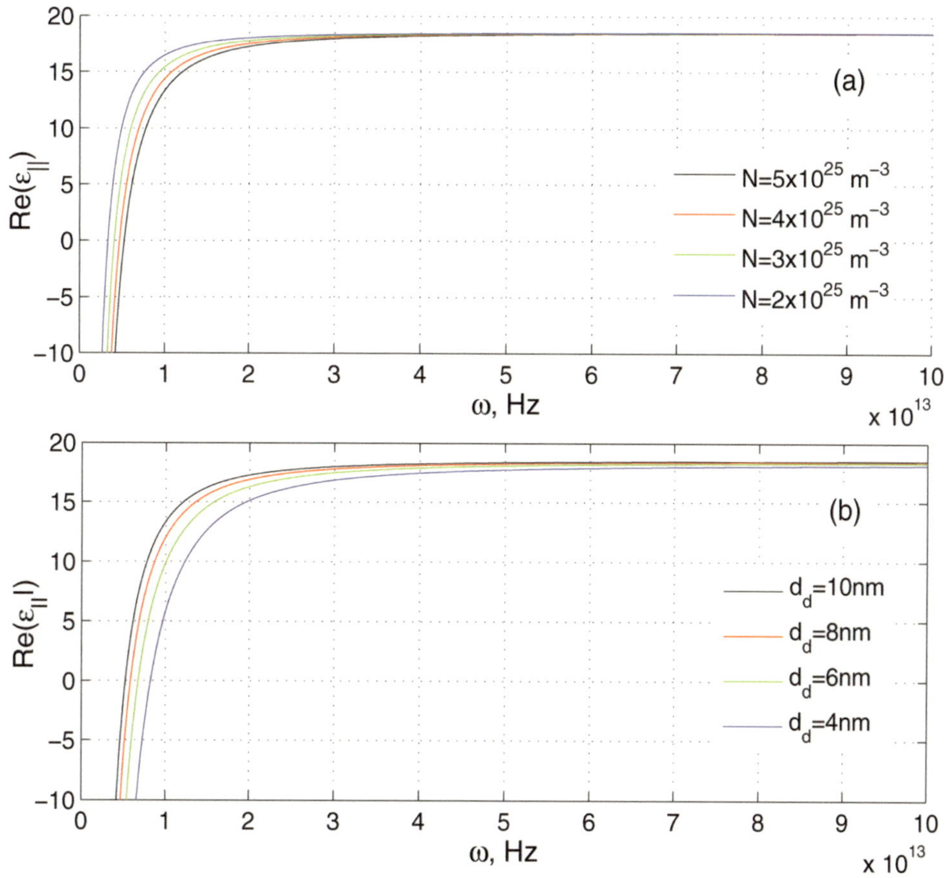

Figure 11. The influence of (a) doping level N and (b) thickness of dielectric d_d, on the real part of $\varepsilon_{\|}$. d_d = 10 nm in (a) and N = 5 × 10^{25} m^{-3} in (b).

Other than the doping level, the resonant behaviour of $\varepsilon_\perp$ was found to depend on the fill factions of the dielectric and semiconductor sheet, as shown in **Figures 9(b)** and **3(b)**. From **Figure 9(b)**, the shift of the resonant frequency of $\varepsilon_\perp$ to the higher frequencies as the thickness d_d is increased can be clearly distinguished.

3.1. The mode structure

Guided by the homogenization of two HMs, the computed dispersion curves are demonstrated. Thus, in **Figure 13**, the dispersion curves for the case $\varepsilon_1 = \varepsilon_3$, $\varepsilon_2 = \varepsilon_4$ and $d_1 \neq d_2 \neq d_3 \neq d_4$ are shown. In this case, we deal with the boundary of two metamaterials with d_1 = 0.35 nm, d_2 = 10 nm, d_3 = 0.25 nm and d_4 = 10.1 nm. Furthermore, due to the great interest in this case, **Figure 13** refers to the dispersion of surface waves with the calculated effective parameters shown in **Figures 9(a), 10(a), 11(a)** and **12(a)**; the blue line is the free-space light line.

The frequency ranges of surface wave can be tuned by changing the doping level of silicon. As it is shown in **Figure 9(a)**, doping level is correlated to the permittivity orthogonal component $\varepsilon_\perp$. Moreover, dispersion curves are shifted to the lower and higher frequencies with the

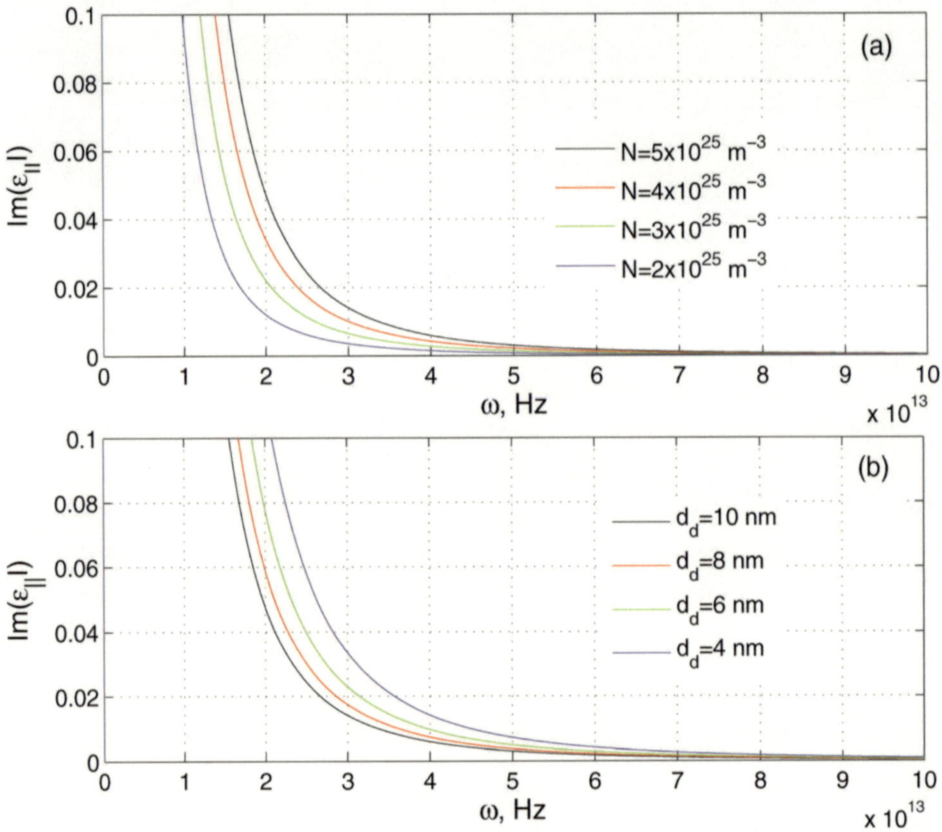

Figure 12. The influence of (a) doping level N and (b) thickness of dielectric d_d, on the imaginary part of $\varepsilon_\parallel$. d_d = 10 nm in (a) and N = 5 × 10^{25} m^{-3} in (b).

http://dx.doi.org/10.5772/intechopen.72801

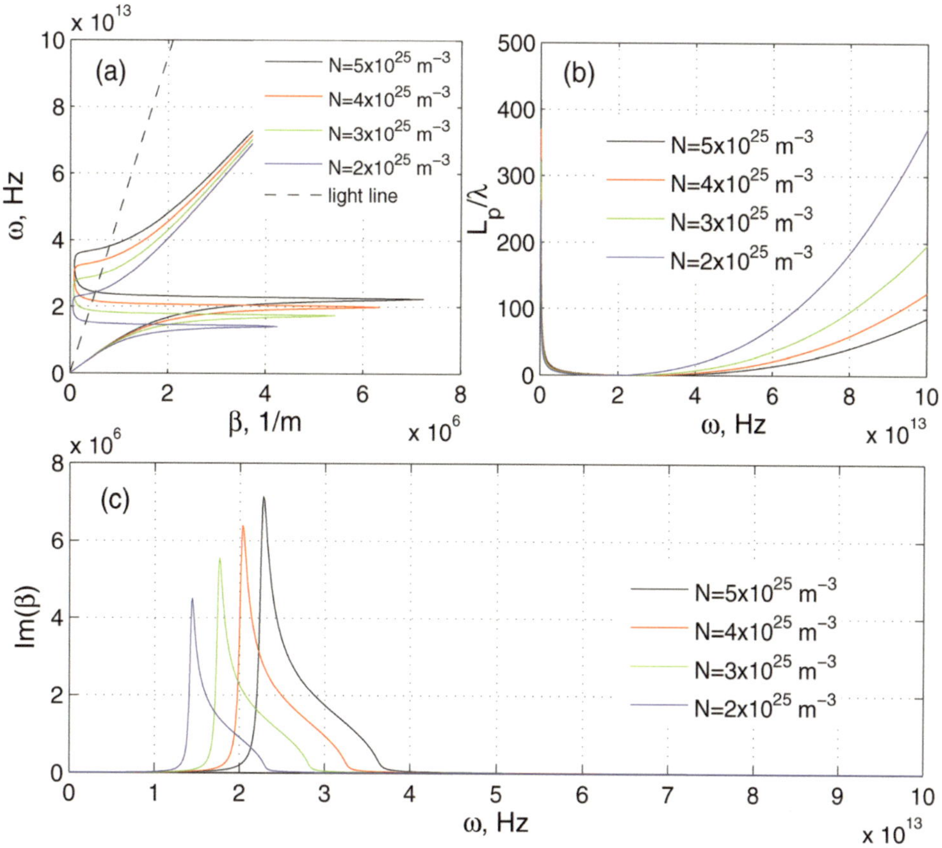

Figure 13. The dispersion of surface waves (a); propagation lengths (b) and absorption (c) at different doping levels of silicon, where $d_d = 10$ nm, $\varepsilon_1 = \varepsilon_3$, $\varepsilon_2 = \varepsilon_4$ and $d_1 \neq d_2 \neq d_3 \neq d_4$ and the blue line in (a) is the free-space light line.

decreases and increases in the doping level N accordingly. These tunability properties can be observed in **Figure 13**. Furthermore, this correlation can be used to engineer the metamaterial surface wave just by controlling silicon sheet doping level.

As the silicon is not modeled as lossless, β is complex, leading to a finite propagation length (Eq. (2)), drawn in **Figure 13(b)**. In **Figure 13** the four modes ($N = 2 \times 10^{25}$ m^{-3}, $N = 3 \times 10^{25}$ m^{-3}, $N = 4 \times 10^{25}$ m^{-3}, $N = 5 \times 10^{25}$ m^{-3}) always lie to the right side of the light line and remain non-radiative (bound) SP modes throughout the certain frequency range. All the considered cases are of particular interest due to the fact that their dispersion relations cross the light line and a significant portion at lower frequencies lies above the free-space light line, which usually splits up non-radiative (bound) and radiative (leaky) regions. For the bound modes, longer propagation lengths take place at lower frequencies owning the dispersion that is close to linear. Mode corresponding to the case $N = 4 \times 10^{25}$ m^{-3} possesses the longer propagation length than the mode corresponding to the case $N = 5 \times 10^{25}$ m^{-3}.

The existence of the boundary modes associated with the second case under consideration, i.e. $\varepsilon_1 = \varepsilon_3$, $\varepsilon_2 \neq \varepsilon_4$ and $d_1 \neq d_2 \neq d_3 \neq d_4$, is also of the particular importance. Thus, **Figure 14**

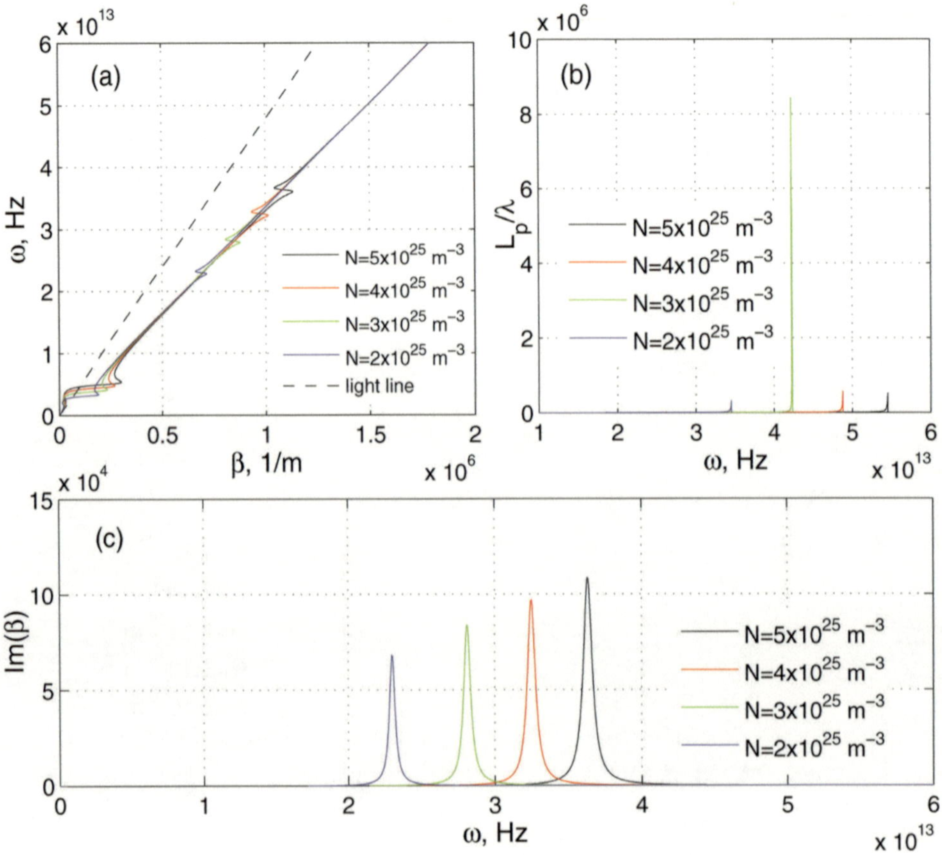

Figure 14. The dispersion of surface waves (a), propagation lengths (b) and absorption (c) at different doping levels, where d_d = 10 nm, $\varepsilon_1 = \varepsilon_3$, $\varepsilon_2 \neq \varepsilon_4$ and $d_1 \neq d_2 \neq d_3 \neq d_4$ and the blue line in (a) is the free-space light line.

tackles this problem by displaying four modes at the boundary of two different metamaterials with ε_4 = 2.25. As seen from **Figure 14**, the smallest asymptotic frequency corresponds to the case N = 2 × 10^{25} m^{-3}.

In contrast to the previous case, we now discuss the instance denoted as $\varepsilon_1 = \varepsilon_3$, $\varepsilon_2 \neq \varepsilon_4$ and $d_1 = d_3$, $d_2 = d_4$ and shown by means of the dispersion diagrams of the TM modes. Thus, **Figure 15** shows the dispersion curves of four different modes. The assessment and control of variation of the dielectric and semiconductor sheets' fill factors is of critical importance (**Figure 15**). First, the impact of the thickness of the dielectric d_d on the dispersion curve (see **Figure 15(a)**) is considered. It is found that the upper limit moves to the higher frequencies as d_d is increased. The former is consistent with the effect of d_d on the frequency range of $\varepsilon_\perp$. The dependence of the frequency range of the surface waves existence on the thickness of dielectric stands for as the most critical feature of the HMs providing an unprecedented degree of freedom to control the surface wave at the near-infrared frequencies. In **Figures 14(a)** and **15(a)**, it is interesting to observe the Ferrell-Berreman modes which exist at energies near the ENZ of the hyperbolic metamaterial to the left of the light line [51–53].

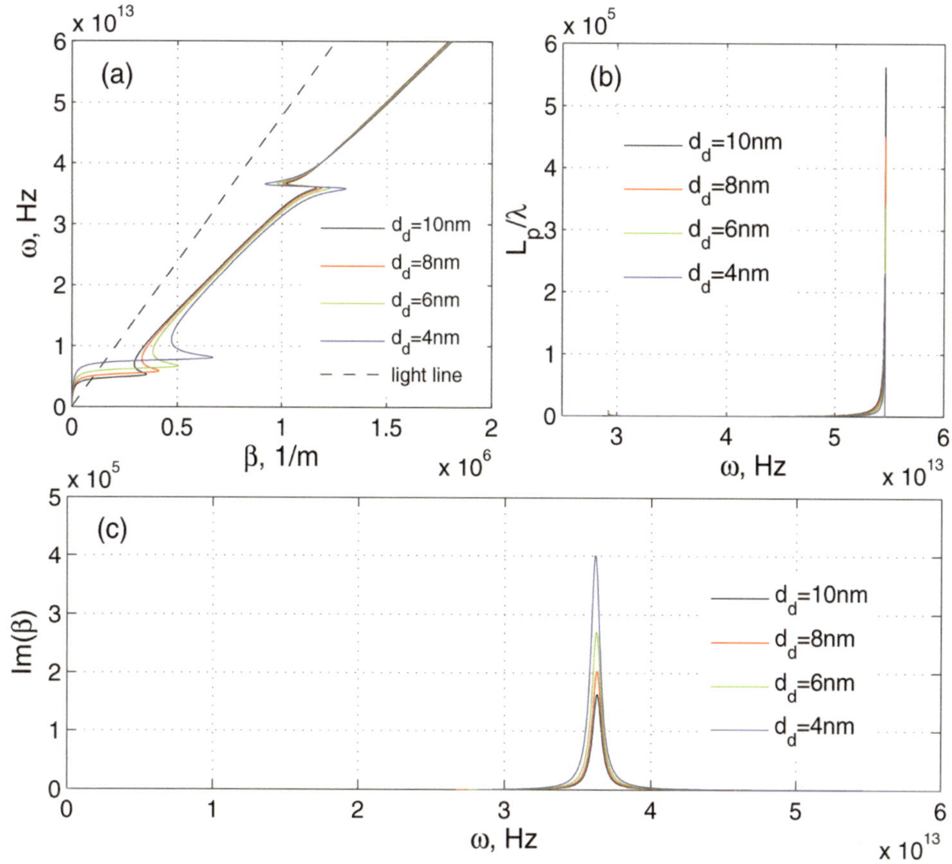

Figure 15. The dependences of dispersion of surface waves, propagation lengths and absorption on (a), (b) and (c) the thickness of dielectric d_d. Herein $N = 5 \times 10^{25}$ m^{-3}.

4. Conclusion

During a study of the HSPPs in a one-dimensional TCO-dielectric MM, we can see that similar to graphene-dielectric MM [45], TCO-dielectric MM supports traditional-like SPPs having different patterns corresponding to two different interfaces. The dispersion equations of HSPPs are obtained based on the theoretical approach [42, 45]. Five kinds of HSPP, among which three kinds are new types of HSPPs and one is the Dyakonov-like SPP and another is the traditional-like SPP have been predicted. The existence of these HSPPs is dramatically influenced by the properties of and the relation among the principal values of the effective permittivity and the dielectric constant of the covering medium. It is worthwhile mentioning that the new types of the HSPPs arise because the principal values of the effective permittivity used in this chapter are functions of frequency and can be negative or positive. Moreover, it was demonstrated that used approach allows to predict surface mode with the dispersion that coincides with the dispersion of a surface plasmon at the boundary of two isotropic media corresponding to the MM interface if the material at the right-hand side is the same as employed in the MM.

Moreover, a new kind of surface wave between two nanostructured semiconductor metamaterials was demonstrated. It is shown that the dispersion diagrams are sensitively dependent on the semiconductor parameters. These findings open the gateway towards potential applications in both classical and quantum optical signal communication and processing.

Author details

Aleksej Trofimov[1]*, Tatjana Gric[1] and Ortwin Hess[2]

*Address all correspondence to: aleksej.trofimov@vgtu.lt

1 Vilnius Gediminas Technical University, Vilnius, Lithuania

2 Imperial College London, London, UK

References

[1] Pendry J. Negative refraction makes a perfect lens. Physical Review Letters. 2000;**85**: 3966-3969

[2] Shalaev V. Optical negative-index metamaterials. Nature Photonics. 2007;**1**:41-48

[3] Shalaev V. Transforming light. Science. 2008;**322**:384-386

[4] Cai W, Shalaev V. Optical Metamaterials: Fundamentals and Applications. New York: Springer Verlag; 2009

[5] Pendry J, Schurig D, Smith D. Controlling electromagnetic fields. Science. 2006;**312**:1780-1782

[6] Schurig D, Mock J, Justice B, Cummer S, Pendry J, Starr A, Smith D. Metamaterial electromagnetic cloak at microwave frequencies. Science. 2006;**314**:977-980

[7] Lal S, Link S, Halas N. Nano-optics from sensing to waveguiding. Nature Photonics. 2007;**1**:641-648

[8] Salandrino A, Engheta N. Far-field subdiffraction optical microscopy using metamaterial crystals: Theory and simulations. Physical Review B. 2006;**74**:75103

[9] Kildishev A, Shalaev V. Engineering space for light via transformation optics. Optics Letters. 2008;**33**:43-45

[10] Tsakmakidis KL, Hermann C, Klaedtke A, Jamois C, Hess O. Surface plasmon polaritons in generalized slab heterostructures with negative permittivity and permeability. Physical Review B. 2006;**73**:085104

[11] Ozbay E. Plasmonics: Merging photonics and electronics at nanoscale dimensions. Science. 2006;**311**:189-193

[12] Bouillard J-S, Vilain S, Dickson W, Zayats A. Hyperspectral imaging with scanning near-field optical microscopy: Applications in plasmonics. Optics Express. 2010;**18**:16513

[13] Dickson W, Wurtz GA, Evans PR, Pollard RJ, Zayats AV. Electronically controlled surface plasmon dispersion and optical transmission through metallic hole arrays using liquid crystal. Nano Letters. 2008;**8**:281

[14] MacDonald KF, Samson ZL, Stockman MI, Zheludev NI. Ultrafast active plasmonics. Nature Photonics. 2009;**3**(1):55-58

[15] Atwater HA, Polman A. Plasmonics for improved photovoltaic devices. Nature Materials. 2010;**9**:205-213

[16] Kabashin A, Evans P, Pastkovsky S, Hendren W, Wurtz G, Atkinson R, Pollard R, Podolskiy V, Zayats A. Plasmonic nanorod metamaterial for biosensing. Nature Materials. 2009;**8**:867-871

[17] Berini P, De Leon I. Surface plasmon–polariton amplifiers and lasers. Nature Photonics. 2011;**6**(1):16-24

[18] Park I-Y, Kim S, Choi J, Lee D-H, Kim Y-J, Kling MF, Stockman MI, Kim S-W. Plasmonic generation of ultrashort extreme-ultraviolet light pulses. Nature Photonics. 2011;**5**(11): 677-681

[19] Liu H, Lalanne P. Microscopic theory of the extraordinary optical transmission. Nature. 2008;**452**(7188):728-731

[20] Kim S, Jin J, Kim Y-J, Park I-Y, Kim Y, Kim S-W. High-harmonic generation by resonant plasmon field enhancement. Nature. 2008;**453**(7196):757-760

[21] Purvis MA, Shlyaptsev VN, Hollinger R, Bargsten C, Pukhov A, Prieto A, Wang Y, Luther BM, Yin L, Wang S, Rocca JJ. Relativistic plasma nanophotonics for ultrahigh energy density physics. Nature Photonics. 2013;**7**(10):796-800

[22] Williams CR, Andrews SR, Maier SA, Fernández-Domínguez AI, Martín-Moreno L, García-Vidal FJ. Highly confined guiding of terahertz surface plasmon polaritons on structured metal surfaces. Nature Photonics. 2008;**2**:175-179

[23] Huang L, Chen X, Bai B, Tan Q, Jin G, et al. Helicity dependent directional surface plasmon polariton excitation using a metasurface with interfacial phase discontinuity. Light: Science & Applications. 2013;**2**:e70

[24] Liscidini M, Sipe JE. Quasiguided surface plasmon excitations in anisotropic materials. Physical Review B. 2010;**81**:115335

[25] Zapata-Rodríguez CJ, Miret JJ, Vuković S, Belić MR. Engineered surface waves in hyperbolic metamaterials. Optics Express. 2013;**21**:19113-19127

[26] Gao J, Lakhtakia A, Lei M. Dyakonov-tamm waves guided by the interface between two structurally chiral materials that differ only in handedness. Physical Review A. 2010;**81**: 013801

[27] Tamm I. Possible type of electron binding on crystal surfaces. Zeitschrift für Physik. 1932;**76**:849-850

[28] Polo JA Jr, Nelatury SR, Lakhtakia A. Controlling hybrid-polarization surface plasmon polaritons in dielectric-transparent conducting oxides metamaterials via their effective properties. Journal of Nanophotonics. 2007;**1**:013501

[29] Vuković SM, Miret JJ, Zapata-Rodríguez CJ, Jakšić Z. Oblique surface waves at an interface between a metal–dielectric superlattice and an isotropic dielectric. Physica Scripta. 2012;**2012**(T149):014041

[30] Zapata-Rodríguez CJ, Miret JJ, Sorni JA, Vuković S. Propagation of dyakonon wavepackets at the boundary of metallodielectric lattices. IEEE Journal of Selected Topics in Quantum Electronics. 2013;**19**:4601408

[31] Salandrino A, Engheta N. Far-field subdiffraction optical microscopy using metamaterial crystals: Theory and simulations. Physical Review B. 2006;**74**:075103

[32] Shvets G, Trendafilov S, Pendry JB, Sarychev A. Guiding, focusing, and sensing on the subwavelength scale using metallic wire arrays. Physical Review Letters. 2007;**99**:053903

[33] Liu Z, Lee H, Xiong Y, Sun C, Zhang X. Far-field optical hyperlens magnifying subdiffraction-limited objects. Science. 2007;**315**:1686

[34] Rho J, Ye Z, Xiong Y, Yin X, Liu Z, et al. Spherical hyperlens for two-dimensional subdiffractional imaging at visible frequencies. Nature Communications. 2010;**1**:143

[35] Naik GV, Shalaev VM, Boltasseva A. Alternative plasmonic materials: Beyond gold and silver. Advanced Materials. 2013;**25**:3264-3294

[36] Cattom MG, Tilley DR. Introduction to Surface and Superlattice Excitations. Bristol: IOP Publishing; 2005, Ch. 8-9

[37] Li R, Cheng C, Ren F-F, Chen J, Fan Y-X, Ding J, Wang H-T. Hybridized surface plasmon polaritons at an interface between a metal and a uniaxial crystal. Applied Physics Letters. 2008;**92**:141115

[38] Sorni JA, Naserpour M, Zapata-Rodriguez CJ, Miret JJ. Dyakonov surface waves in lossy metamaterials. Optics Communications. 2015;**355**:251

[39] Takayama O, Artigas D, Torner L. Practical dyakonons. Optics Letters. 2012;**37**:4311

[40] Orlov AA, Voroshilov PM, Belov PA, Kivshar YS. Engineered optical nonlocality in nanostructured metamaterials. Physical Review B. 2011;**84**:045424

[41] Zhang Y-L, Zhang Q, Wang X-Z. Extraordinary surface polaritons in obliquely truncated dielectric/metal metamaterials. Journal of the Optical Society of America B. 2016;**33**:543

[42] Iorsh I, Orlov A, Belov P, Kivshar Y. Interface modes in nanostructured metal-dielectric metamaterials. Applied Physics Letters. 2011;**99**(15):151914

[43] Ling RT, Scholler JD, Ufimtsev PY. The propagation and excitation of surface waves in an absorbing layer. Progress in Electromagnetics Research. 1998;**19**:49-91

[44] Campione S, Marquier F, Hugonin J-P, Ellis AR, Klem JF, Sinclair MB, Luk TS. Directional and monochromatic thermal emitter from epsilon-near-zero conditions in semiconductor hyperbolic metamaterials. Scientific Reports. 2016;**6**:34746

[45] Gric T, Hess O. Tunable surface waves at the interface separating different graphene-dielectric composite hyperbolic metamaterials. Optics Express. 2017;**25**(10):11466-11476

[46] Li S, Jadidi MM, Murphy TE, Kumar G. Terahertz surface plasmon polaritons on a semiconductor surface structured with periodic V-grooves. Optics Express. 2013;**21**:7041-7049

[47] Kumar G, Li S, Jadidi MM, Murphy TE. Terahertz surface plasmon waveguide based on a one-dimensional array of silicon pillars. New Journal of Physics. 2013;**15**:085031

[48] Rusina A, Durach M, Nelson KA, Stockman MI. Nanoconcentration of terahertz radiation in plasmonic waveguides. Optics Express. 2008;**16**:18576-18589

[49] Shen L, Chen X, Yang T-J. Terahertz surface plasmon polaritons on periodically corrugated metal surfaces. Optics Express. 2008;**16**:3326-3333

[50] Gric T. Surface-plasmon-polaritons at the interface of nanostructured metamaterials. Progress in Electromagnetics Research. 2016;**46**:165-172

[51] Vassant S, Hugonin J-P, Marquier F, Greffet J-J. Berreman mode and epsilon near zero mode. Optics Express. 2012;**20**:23971-23977

[52] Vassant S, Archambault A, Marquier F, Pardo F, Gennser U, Cavanna A, Pelouard J, Greffet J-J. Epsilon-near-zero mode for active optoelectronic devices. Physical Review Letters. 2012;**109**:237401

[53] Campione S, Brener I, Marquier F. Theory of epsilon-near-zero modes in ultrathin films. Physical Review B. 2015;**91**:121408

Chapter 2

Understanding the Mechanisms that Affect the Quality of Electrochemically Grown Semiconducting Nanowires

Abhay Singh and Usha Philipose

Additional information is available at the end of the chapter

http://dx.doi.org/10.5772/intechopen.71631

Abstract

Template-assisted synthesis of nanowires is a simple electrochemical technique commonly used in the fabrication of semiconducting nanowires. It is an easy and cost-effective approach compared to conventional lithography, which requires expensive equipment. The focus of this chapter is on the various mechanisms involving mass transport of ions during successive stages of the template-assisted electrochemical growth of indium antimonide (InSb) nanowires. The nanowires were grown in two different templates such as commercially available anodic aluminum oxide (AAO) templates and polycarbonate membranes. The chapter also presents the results of characterizing the InSb nanowires connected in a field effect transistor (FET) configuration. The Sb-rich InSb nanowires that were fabricated by DC electrodeposition in nanoporous AAO exhibited hole-dominated transport (p-type conduction). Temperature-dependent transport measurement shows the semiconducting nature of these nanowires.

Keywords: semiconducting nanowire, InSb, electrodeposition, anodic aluminum oxide (AAO), nanowire growth mechanism

1. Introduction

In recent years, anodic aluminum oxide (AAO) templates have been used extensively for fabricating myriads of nanostructures. AAO is a self-organized nanostructured material containing a high density of uniform cylindrical pores that are aligned perpendicular to the surface and extend over the entire thickness of the template. AAO is also optically transparent, electrically insulating, thermally and mechanically robust and chemically inert. Nanowire arrays that are grown in AAO template pores are used in many applications such as energy conversion [1], energy-storage devices [2, 3], electronics [4], metamaterials [5], optoelectronics [6], photonics [7], and piezoelectrics [8]. Template-assisted electrochemical growth of large-scale nanowire arrays is attractive because they are readily scalable to mass production. Moreover, it is possible to obtain an array of uniform diameter and length and a surface that can be engineered by nonuniformity of the template tubes.

This chapter reviews the various processes that govern the filling of the nanochannel alumina template pores resulting in a systematic growth of nanowires within the pores. We also discuss the challenges faced in achieving uniform growth including effects of pore-wetting and aeration. Any nonuniformity in the alumina template tubes also result in nanowires with rough surfaces, an exploitable result for use of nanowires in thermoelectric applications.

Over the last decade, there has been increasing interest in nanostructures of III–V semiconducting materials due to their potential applications in electronic and optoelectronic devices [9]. After identifying the various mechanisms of electrochemical growth of semiconducting nanowires, we present the specific growth process conditions for growth of indium antimonide (InSb) nanowires. Indium antimonide (InSb) bulk is a promising III–V direct bandgap semiconductor material with zinc blende (FCC) structure [10–14]. InSb has a high room temperature mobility (electron and hole mobility [15] of 77,000 and 850 $cm^2 V^{-1} s^{-1}$, respectively), low electron effective mass [16] of 0.014, and low direct bandgap (E_g = 0.17 eV, at 300 K), and large Lande g-factor of 51, [17] making it suitable for use in applications such as high speed, low-power transistors, tunneling field effect transistors (FETs), infrared optoelectronics [18], and magnetoresistive sensors [19].

Nanowires can be grown via electrodeposition in different nanoporous templates like anodic aluminum oxide (AAO) and ion track-etched polycarbonate. Every template type has its own advantage, disadvantage, and application.

2. Growth process: convection, migration, and diffusion

The benchmark method for synthesis of nanowire was suggested in 1986 by Martin et al. [20], where nanowires were grown inside polycarbonate membranes. This method has since been widely used to synthesize nanowires in both polycarbonate and anodic aluminum oxide (AAO) membranes. These membranes contain a high density of cylindrical pores that are perpendicular to the membrane surface and in most cases, these pores penetrate the entire membrane thickness. There have been several studies on electrochemical growth of nanowires but very small number of investigations on understanding the growth mechanism of nanowires in AAO pores.

Template-assisted nanowire growth process consists of two major steps: electrochemical reduction of the cation inside the pore, followed by removal of the template. Several studies [21, 22] have shown that during the pore-filling stage (nanowire growth), the total constant current density is independent of flux variation inside the pore and concentration of electrolyte at the mouth of pore during electrodeposition. Recently, another study [23] showed that the electrolyte concentration at the mouth of the pore remains constant and current decays with time; a fact that contradicts the previous study where current remains constant. Here, we discuss the more realistic model that describes a situation where the electrolyte concentration at the mouth of the pores is different than the bulk during electrodeposition.

The electrodeposition growth process of nanowires in porous templates has been modeled by considering different mechanisms of mass transport at different stages of the growth process [24]. There are mainly three different mechanisms that could explain the growth process: (i) diffusion process (arising from concentration gradients existing between electrolyte in the template pores and in the solution), (ii) convection process (arising from movement of deposition ions in the electrolyte), and (iii) migration of ions into the template pores. Contribution of each process can be controlled by modifying the cell design, reducing depositing ion concentrations compared to mixing electrolyte, and stirring the electrolyte during growth process to avoid the concentration gradient.

2.1. Diffusion mechanism in nanowire growth process

Mass transport of ions during nanowire growth is mainly controlled by a diffusion process. The three main stages of this process are shown in **Figure 1**. **Figure 1(a)** shows the earlier stage of diffusion where diffusion front propagates in only one dimension inside the pores.

Figure 1(b) shows the second stage where diffusion front reached the mouth of the pore and opens in three-dimensional hemispherical front. As soon as diffusion front reaches at the mouth, then there is a concentration gradient builds up between the mouth and the bulk. **Figure 1(c)** shows the third stage where linear diffusion happens inside and outside of the pores. This diffusion-limited electrodeposition results in a morphological instability driven by individual nanowires [25]. In template-assisted electrodeposition, nanowire length must follow the length of AAO pore, but some nanowire grows fast compared to others and ends up mushroom-like structures at the top surface of AAO template [26–31]. Recent study [25] also showed that this growth instability can be avoided by introducing temperature gradient between bottom and top of the AAO pores.

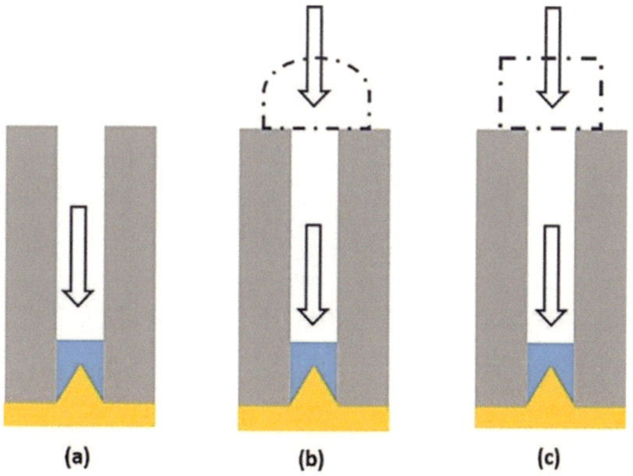

Figure 1. Schematic of growth of nanowire in a single nanopore: (a) linear diffusion, (b) linear and semi-infinite hemispherical diffusion, and (c) semi-infinite planar diffusion outside the pore (Adapted from Ref. [24]).

The diffusion mechanism during various stages of nanowire growth is described by the S-curve of current-time plot, as shown in **Figure 2(a)**.

The entire mechanism involves three consecutive growth stages: (i) one dimensional diffusion front inside AAO pores; (ii) linear diffusion inside the pores and hemispherical diffusion at the mouth of the pores; and (iii) merging of hemispherical diffusion front and resulting in planned front over the surface of the mouth. The electrodeposition time and average length of wire have been controlled by monitoring the deposition current versus time at given deposition voltage −1.5 V which was maintained with respect to reference electrode. Inset of **Figure 4(a)** shows the schematic of the various stages of electrodeposition.

In first stage of growth process, concentration of the electrolyte at the mouth of pores matches the concentration of the bulk of the electrolyte (c^b). The linear diffusion front passing through pore length results in transportation of In and Sb ions inside the pores of the AAO template. The diffusion fronts between individual pores have not been overlapped at this stage. The concentration within the pore changes as nanowire growth starts and it has been expressed by Fick's law as: $c(x,t) = c^b \mathrm{erf}\left[\frac{x}{2\sqrt{Dt}}\right]$, where $2\sqrt{Dt}$ is the diffusion length (L) that corresponds how far concentration varies along the pore length at time t. The initial pore length is L_0 at $t = 0$. There is a sharp decrease in electrodeposition current during this time, as is expected for diffusion-controlled process. The current was observed to decrease from 35 to 20 mA in region (I) of the plot in **Figure 2(a)** which corresponds to region (i) of growth process. In region (i), In and Sb ions diffuse to the bottom of the pore in the tubular column of AAO template to initiate nanowire growth.

During region (II) of the growth process, initial linear diffusion front reaches the mouth of the pores and three dimensional hemispherical diffusion front develops at the mouth. During this stage in region (II) of plot **Figure 4**, the steady increase in current has been observed and is believed to increase the diffusion of ions to the nanowire growth front. During the final stage in region (III), the nanowires have filled the pore completely and the hemispherical cap merges with each other and formed a continuous rough film on the surface of template. In the

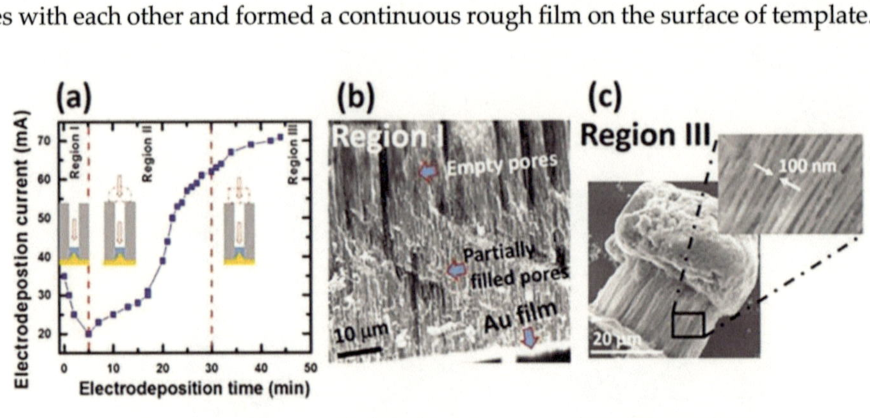

Figure 2. Evolution of nanowire growth in the AAO template pores: (a) current-time plot of InSb nanowire growth showing different growth regions; (b) cross-sectional SEM image of the template in region I. The thin bottom layer is a gold film. The pores are partially filled with InSb. This region corresponds to a planar diffusion inside the pore; and (c) cross-sectional SEM image of region III where NWs have filled the pores and the dome-shaped tips collapse to form mushroom-shaped clusters, corresponding to semi-infinite planar diffusion at the mouth of the pore [32].

http://dx.doi.org/10.5772/intechopen.71631

region (III) of the S-curve plot, current does not increase significantly and tends to saturate, as compared to region (II) of the plot. **Figure 2(b)** and **(c)** shows cross-sectional SEM images of template in the region (I) and (III), respectively. **Figure 2(b)** shows partially filled pores, which correspond to a linear diffusion in region (I) of the S-curve. **Figure 4(c)** shows the completely filled pores and hemispherical caps that merge to form a dense mushroom-like overgrowth on the surface of the pores.

2.2. Convection mechanism in growth of nanowires

The effect of convection on nanowire growth process is dependent on cell design. The three electrodes in an electrochemical cell are: (i) working electrode (substrate); (ii) counter electrode (Pt wire/mesh); and (iii) reference electrode. To minimize the mechanism of natural convection, electrodeposition cell configuration is an integral part of nanowire growth in nanoporous templates. In **Figure 3(a)** and **(b)**, the cathode and anode are placed vertically above each other, whereas in **Figure 3(c)**, the cathode and anode are placed parallel to each other. The effects of convection are considered to be minimal for the design in **Figure 3(c)**. Konishi et al. [33] reported that the electrodeposition current increases in the case of anode over cathode **Figure 3(b)** configuration, and current decreases in the case of cathode over anode configuration **Figure 3(a)** during early stages of the electrodeposition. They have found the considerable difference in the transient behavior of the current in 100- and 200-nm-sized pores in certain stage of electrodeposition, which is caused by the electrolytic cell configuration.

The effects of natural convection are proven to be significant and reported in Ref [32] for the schematic shown in **Figure 4**, where an increase in convection current because of this mechanism is shown for pores size as small as 100 nm. This effect was observed during Cu electrodeposition.

2.3. Migration of ions during growth of nanowires

To explain the InSb nanowire growth process in the following section, the effect of migration of In and Sb ions has not been considered in this discussion. Since concentration of these ions in supporting electrolyte is low, the contribution of ion migration is considered to be lower than diffusion.

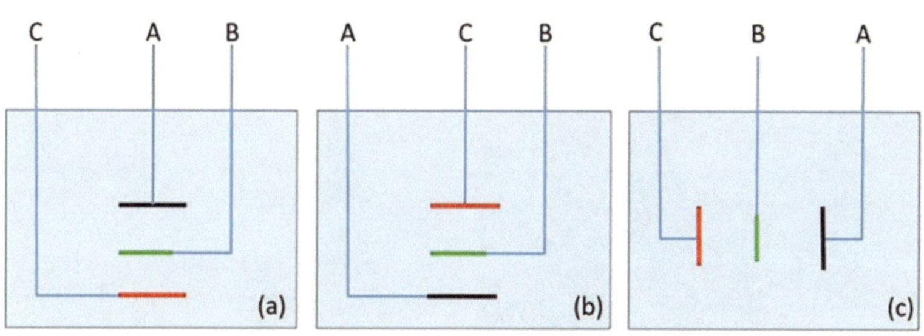

Figure 3. Schematic of cell designs: (a) cathode over anode; (b) anode over cathode; and (c) cathode and electrode placed laterally and parallel to each other. Letter A represents working electrode (Cathode), C represents secondary electrode (Anode), and B represents reference electrode.

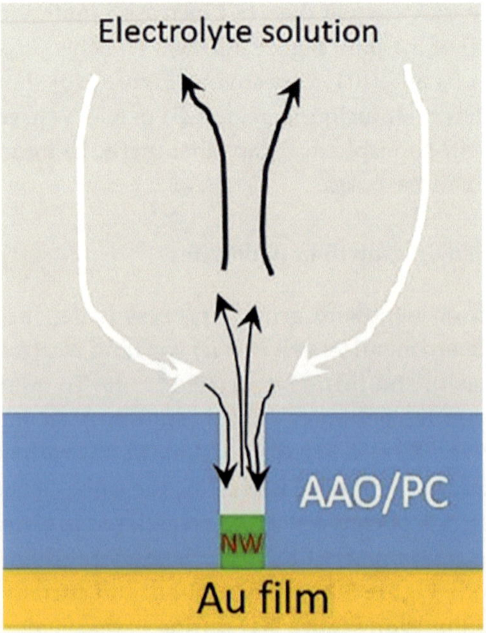

Figure 4. Schematic diagram convection in and around nanoporous pores in the configuration of anode over cathode.

To conclude the discussion on mechanisms affecting nanowire growth in template pores, we can say that there are various parameters that affect the nanowire growth such as cell design, pH of electrolyte, applied bias as well as various growth mechanisms like convection, migration and diffusion. However, of these, diffusion-controlled mass transfer flux is more important. Influence of migration can be ignored by lowering the concentration of depositing metal and increasing the concentration of supporting electrolyte. The contribution due to natural convection can be avoided by electrodeposition cell design in which working electrode can be placed in up or vertical positions.

Commercial self-organized porous AAO templates purchased from Synkera Technologies, Inc., have AAO pore length and diameter of ~50 μm and ~100 nm, respectively. **Figure 5** shows the

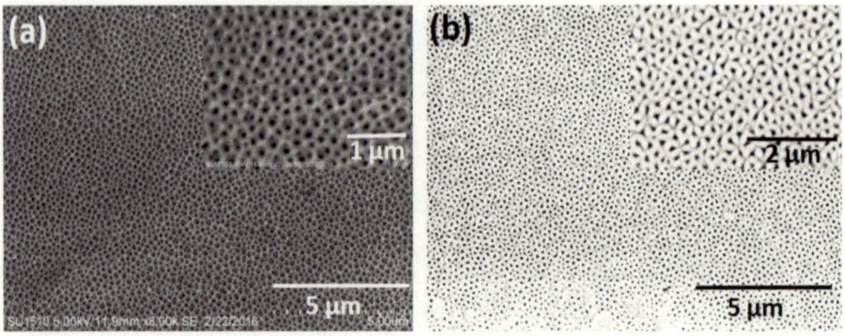

Figure 5. SEM image of commercial anodic aluminum oxide (AAO) template purchased from Synkera Technologies Inc. (a) SEM image of top surface of AAO template at 500-nm scale bar and inset shows 1-μm scale bar. These pores are nonuniform pores. (b) SEM image of bottom surface of AAO template at 500-nm scale bar.

SEM images of the (a) top and (b) bottom surfaces of commercial AAO templates. Inset shows higher magnification of the same surfaces. It is obvious from these images that the top and bottom pores do not have the same diameter, and the top surface pores are larger in diameter compared to the bottom pores. It is this difference in pore diameters that leads to most researchers depositing gold on the lower diameter side of the AAO template, thus preventing overfilling of the pores.

3. Fabrication and characterization of InSb nanowires in commercial AAO template

Using the commercial AAO templates discussed in the previous section, one surface of the AAO template was coated with a thin film of Au (~150 nm) using thermal evaporation. Prior to Au evaporation, the pores were widened nominally by soaking the template in 5% phosphoric acid (H3PO3) for 2 min at 30°C. The electrodeposition process was conducted in a three-electrode cell with the AAO template Au side as the working electrode, platinum electrode as the counter electrode and Ag/AgCl as the reference electrode. Crucial part of deposition process was to cover the edges of the AAO template with an insulating material and then attaching it to a conducting copper tape. The side of the AAO pore, which was covered with thin film of Au, was attached to conducting glue side of copper tape. After that, insulating polymer was applied to avoid any conducting path except through the open AAO pores which terminates at the Au film at the bottom of the pores. The entire electrodeposition process was controlled and carried out by a potentiostat (Princeton Applied Research, model: Potentiostat/Galvanostat 263A). The deposition parameters used for electrolyte in this experiment are as follows: 0.15 M $InCl_3$, 0.1 M $SbCl_3$, 0.36 M $C_6H_8O_7H_2O$, and $C_6H_5Na3O_7$. The pH value of the electrolyte was adjusted to 1.7 and electrodeposition was performed at room temperature (200°C). The citrate ions have been used as complexing agents that bring the deposition potential of In and Sb closer to maintain a binary growth during the deposition of the InSb nanowires. The complete electrodeposition process was performed for 50 min at a potential of −1.5 V with respect to the reference electrode (Ag/AgCl). Also, magnetic stirring was used to maintain the uniform concentration of electrolyte during growth of nanowires. On completion of the deposition process, the AAO template was carefully rinsed several times with DI water, and then AAO template was removed from copper tape for further processing.

Following steps are involved in extraction of nanowires from AAO template: (a) AAO template was placed in a clean 80 ml beaker with the gold side facing upward and gold film was removed by 1 M potassium iodide (KI). To dissolve the AAM template, a few drops of 1 M KOH solution was used. This was done with vigorous shaking for about 20 min (b) following dissolution of the template; 15 ml of DI water was added to the solution, which was then sonicated for 10 min. The sonication step helps to dissolve the AAO completely, while breaking the InSb nanowires off the InSb crust (overgrown areas). (c) DI water as added to wash out the precipitates in the bottom of the beaker into the vial; the process was repeated until there was no visible precipitate in the beaker. This step was extremely important, since most of the nanowires were obtained from these precipitates. (d) To further dilute the KOH and AAO in the solution, both vials were placed in a centrifuge (Fisher Scientific Centrifuge, model: 228

Benchtop Centrifuge) at 3000 rpm for 10 min. Eighty percent of the solution was then carefully removed from the top of the centrifuge vial. Dilution step was repeated until there was no residue of KOH and AAO left, a fact that was verified by SEM. The final dark solution was found to contain a high density of InSb nanowires that could then be drop-cast for making devices. After growth, overgrown part is removed by a gentle mechanical polishing. Average length and diameter of InSb nanowire after processing have been found to be 20–30 μm and 100 nm, respectively.

Figure 6(a) is an SEM image of dispersed InSb nanowires, which show that they have a rough surface. This surface roughness is attributed to nonuniform pores and roughness inside the pores of the AAO template. This roughness can be avoided by using homemade AAO template or track-etched polycarbonate membranes where the tubular pores are uniform resulting in smooth and uniform nanowires as compared to those grown in commercially available AAO templates. The nanowire composition was verified by energy dispersive X-ray spectroscopy (EDX) in **Figure 6(b)**, which shows Sb-rich composition, with an average In:Sb weight ratio of 40:60.

To verify the composition and crystallinity, Raman spectroscopy and X-ray diffraction studies were made on the as-grown InSb nanowires. **Figure 7** shows the Raman spectrum obtained from two crossed InSb nanowires that were dispersed on a Si substrate. Optical image of the region where the spectrum was collected from is shown in the inset of **Figure 7**. Room temperature Raman spectrum shows two distinct peaks at 178 and 188 cm^{-1}, which correspond to TO and LO phonon modes, respectively. These peaks are matched to prior studies on InSb nanowire [9]. Additional peaks at 150 (TO-TA) and 110 cm^{-1} have been reported earlier in InSb nanorods by Wada et al. [34].

Above InSb NWs was characterized with high-resolution X-ray diffraction system from Rigaku model Ultima III. **Figure 8** shows the X-ray diffraction spectrum of as-grown InSb nanowires in AAO template. All strong peaks were indexed to 2θ = 23.0, 39.0, and 46 identify

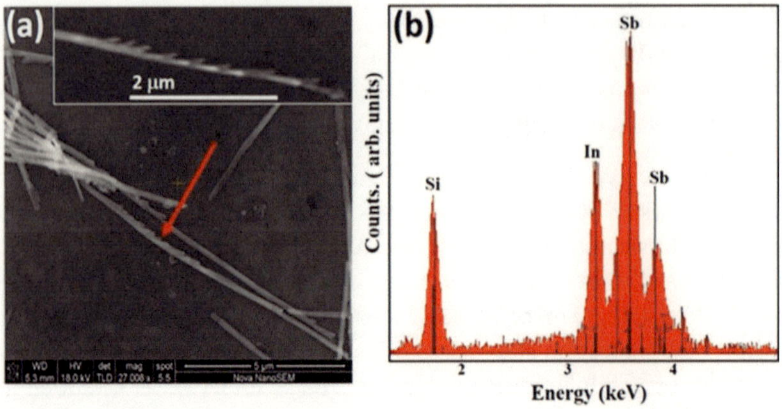

Figure 6. SEM images of (a) InSb nanowires that were removed from the AAO template and dispersed on a cleaned Si substrate. Inset shows a single nanowire with roughened surface and (b) EDX spectrum of Sb-rich InSb nanowires [32].

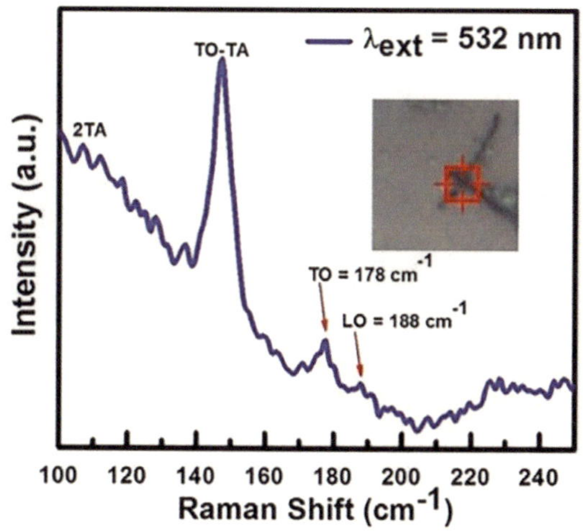

Figure 7. Raman spectrum obtained from two crossed InSb nanowires; inset shows the region from where the Raman spectrum was collected. The two peaks at 150 and 110 cm^{-1} are most likely related to surface roughness and defects in the nanowire. The characteristic TO and LO peaks attest to crystallinity of the InSb nanowires [32].

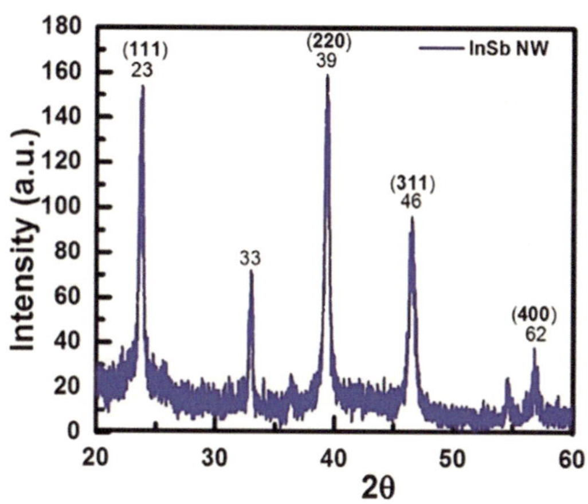

Figure 8. XRD spectrum of InSb NWs grown in AAO template in the scan range of 2θ = 20–60°.

the dominant phase as zinc blende InSb. No other crystalline impurities peaks, such as In2O3, were detected in the XRD pattern. Similar XRD pattern has been reported earlier for electrochemically grown InSb nanowires [35–37].

To obtain additional information on the electronic quality of the as-grown InSb nanowires, electron transport measurement was performed on individual nanowires by connecting them in a back-gated FET type structure as shown in **Figure 9**. To determine type, concentration and mobility of carriers in the nanowire, 2-terminal and 3-terminal current–voltage measurements were performed on the fabricated device. **Figure 9(a)** shows the device schematic where back-gate has been used and **Figure 9(b)** shows the SEM image of the InSb nanowire contacted by Cr/Au as a source and drain.

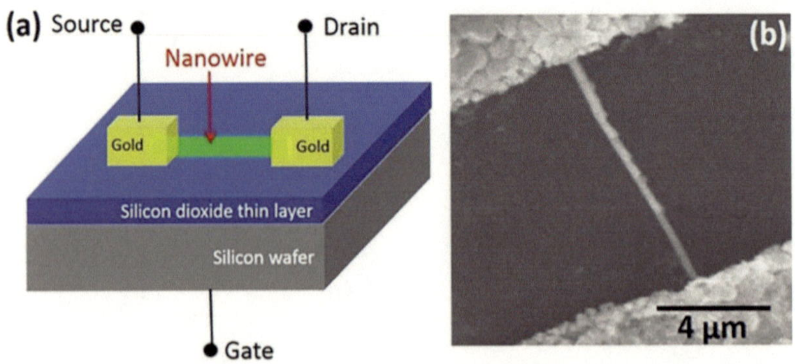

Figure 9. (a) Schematic of back-gated InSb nanowire field effect transistor and (b) SEM image of a single InSb nanowire contacted by two Cr/Au contacts [32].

Figure 10 shows the results of electron transport measurements made on the back-gated nanowire at different V_{ds} (from 0.2 to 1.0 V, in steps of 0.2 V). The transfer characteristic at V_{ds} = 1.0 V was found to have a subthreshold region between −0.5 and +0.5 V in both linear and log scale (Inset of **Figure 10.**). These values most likely correspond to complete depletion of the nanowire. The conductance of InSb nanowire has been shown to decrease with the increasing V_g, that is, confirmation of a p-channel FET behavior. The p-type behavior is attributed to the Sb-rich nature of the InSb nanowire and attributed to the two common defects in Sb-rich InSb and Sb antisite and In interstitial defects [38].

The equilibrium hole concentration p_0 in the nanowire was determined by using the Eq. (1) [39, 40]:

$$p_0 = \frac{C V_T}{q\pi R^2 L} \tag{1}$$

where the gate capacitance (C) has been determined from the Eq. (2) [40]:

$$C = \frac{2\pi \varepsilon_0 \epsilon_{eff} L}{cosh^{-1}\left(\frac{t_{ox} + R}{R}\right)} \tag{2}$$

where ϵ_{eff} ~ 2.2 is the effective dielectric constant of the SiO2 dielectric, ϵ_0 is the permittivity of free space, t_{ox} = 250 nm is the thickness of dielectric, R = 50 nm is the nanowire radius, and L = 10 μm is the length of nanowire channel, which is used for this measurement. Using Eqs. (1) and (2), the hole concentration has been calculated to be 1.96×10^{16} cm^{-3}. From the linear part of the I_{ds} versus V_{gs} as shown in **Figure 10**, the field effect mobility (μFE) was determined at different constant V_{ds} using Eq. (3):

$$\mu_{FE} = \left(\frac{L^2}{C V_{ds}}\right) \times \left(\frac{d I_{ds}}{d V_{gs}}\right) \tag{3}$$

http://dx.doi.org/10.5772/intechopen.71631

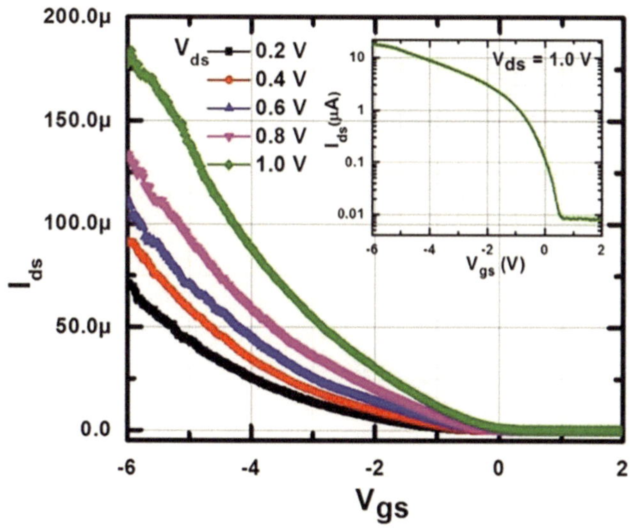

Figure 10. Transfer characteristics of InSb nanowire back-gated FET at increasing V_{ds} from 0.2 to 1.0 V. Inset shows a logarithm plot of the I_{ds} *vs.* V_{gs} curve at $V_{ds} = 1.0$ V. The ON–OFF current ratio is estimated to be of the order of 10^3 [32].

where transconductance $g_m = \left(\frac{dI_{ds}}{dV_{gs}}\right)$ has been deduced at different constant Vds. The field effect mobility was determined to be 507 cm^2 V^{-1} s^{-1} at $V_{ds} = 1.0$ V. The lowest mobility has been found to be 277 cm^2 V^{-1} s^{-1} at $V_{ds} = 0.2$ V, which is higher than the previously reported [11] value of 57 cm^2 V^{-1} s^{-1} obtained at $V_{ds} = 0.1$ V on unintentionally doped 5-μm long p-type InSb nanowire grown by electrochemical method. Similarly, carbon-doped p-type InSb nanowire of length 1.8 μm has been reported to be has the mobility of 127 ± 21 cm^2 V^{-1} s^{-1} at $V_{ds} = 0.05$ V [41]. Inset of **Figure 10** shows p-type nanowire FET has been shown a relatively high ON/OFF ratio of 10^3. One possible cause for the Sb-rich nature of these nanowires is the lower pH (1.7) of the electrolyte. This most likely causes an increase in adsorption of Sb anions on the growing crystalline nanowire.

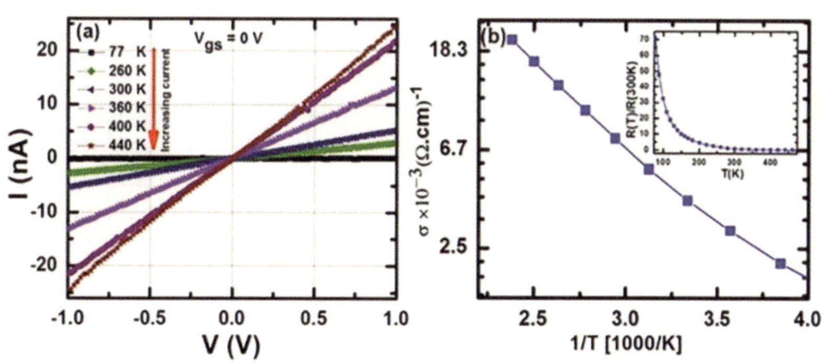

Figure 11. (a) Temperature-dependent I-V shows current increasing with temperature, which is characteristic semiconducting behavior. (b) Temperature-dependent conductivity measurements (Arrhenius plot) for the extraction of activation energy from a single InSb nanowire. Inset shows normalized resistance versus temperature, which shows the exponential decrease of resistance with increasing temperature [32].

The results of temperature-dependent I-V have been shown in **Figure 11(a)**. Linear I-V shows that the Cr/Au contacts (source-drain) to the nanowire are ohmic contacts. The InSb nanowire resistivity was determined to be 248 Ω cm^{-1}, which is higher than the value reported for bulk InSb. The reasons for higher resistivity of the nanowire compared to bulk are as follows: (i) Sb antisite and In interstitials and (ii) surface roughness of nanowire (**Figure 6(a)**). The increased resistivity has been also reported in the references [15, 42], attributed to a significantly reduced hole mobility caused by the scattering at nanowire surface. The nanowire resistance was found to decrease with increasing temperature (inset of **Figure 11(b)**). This large temperature dependence of resistance of InSb nanowire has been reported previously in Ref.s [15, 43]; this is the characteristic behavior of a semiconductor where carrier concentration varies exponentially with temperature. The nanowire conductivity has been measured using device geometry. **Figure 11(b)** shows that below T = 200 K, the nanowire conductivity (σ) exhibits the characteristics of thermal activation which can be analyzed using Eq. (4):

$$\sigma = \sigma_0 exp\left(\frac{-E_a}{k_B T}\right) \tag{4}$$

where σ_0 is the pre-exponent factor, E_a is the activation energy, k_B is Boltzmann's constant (8.617×10^{-5} eVK^{-1}), and T is the absolute temperature. E_a has been estimated to be about 0.1 eV corresponding to carrier generation across the bandgap with activation energy equal to half of the bandgap. The roughness of the nanowire surface can be exploited to control phonon transport. Future work can be directed toward reducing the intrinsic point defects within the nanowire so that electrical conductivity can be enhanced, while the surface roughness can impede transport of phonons to reduce the lattice contribution to thermal conductivity, making it a promising material for thermoelectric applications.

4. Homemade AAO template fabrication for semiconducting nanowire growth of desired dimensions

Anodic films on aluminum have received considerable attention due to their extensive application as templates for synthesizing various nanostructures in the forms of nanowires and nanotubes [44–46]. To obtain a higher control of the nanowire dimensions and surface quality, an electrochemical self-assembly technique was used to fabricate a hexagonally ordered array of cylindrical nanopores on an aluminum substrate. Starting with high purity, unpolished and annealed aluminum (99.997%, AlfaAsar) foil with thickness of ~250 μm, the unpolished samples were first chemically polished [47] using 15 parts of 68% nitric acid and 85 parts of 85% phosphoric acid for 5 min at 850°C. The samples were then neutralized in 1 M sodium hydroxide (NaOH) for 20 min. This was followed by a multistep anodization process using 3% oxalic acid and 40 V DC at room temperature. Final step of anodization is carried out for 5 min, which produces pore length of ~1 μm [48]. **Figure 12** shows SEM images of top surface of AAO template after second anodization, **Figure 12(a)** is the SEM images of top surface at a scale bar of 500 nm and **Figure 12(b)** is the top surface with a scale bar which shows the pore diameter of AAO template is approximately 50 nm.

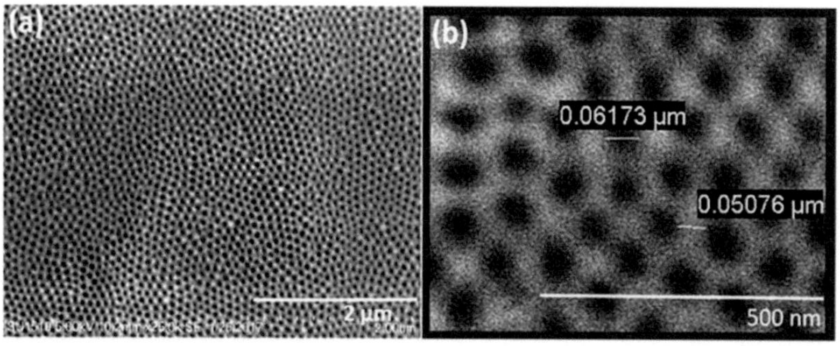

Figure 12. SEM images of AAO template top surfaces (a) at scale bar of 500 nm and (b) with scale bar in pore diameter which shows that diameter of pore is approximately 50 nm.

After second step anodization, aluminum is dissolved in 1 M mercury chloride (H_gCl_2) saturated solution. We have achieved ~50 μm pore length in about 4 h 16 min with pore diameter of ~50 nm. One advantage for the home-grown route of synthesizing templates is the possibility of obtaining smaller pore diameters. In this chapter, results of pore widths of ~50–70 nm are presented, in contrast to the larger diameter ~100 nm obtained from the commercial templates. Nanowires of different diameters will enable a study of its dependence on the electrical, magnetic, and thermoelectric properties of nanowires.

A noteworthy challenge in homemade AAO templates is the presence of a thin barrier layer comprising aluminum oxide. Following fabrication of the alumina layer and removal of the underlying metallic aluminum, the bottom layer of the template was found to exhibit bulges or protrusions. **Figure 13(a)** shows an SEM image showing the capped protrusions (hemispherical caps) of the barrier later at the bottom of the pore. Inset shows an enlarged image of the barrier caps. **Figure 16(b)** shows the completely opened barrier later after etching in 5% phosphoric acid at 30°C for 30min. It has also been observed that if etching time in 5% phosphoric acid is increased from 30min to 35 min, the thin membrane tears in sections from the top surface and the pores merge together as their walls collapse. Such membranes cannot be used for nanowire growth via electrodeposition, since the collapsed walls will result in direct contact with the underlying metal layer. **Figure 14** shows SEM images of bottom and top

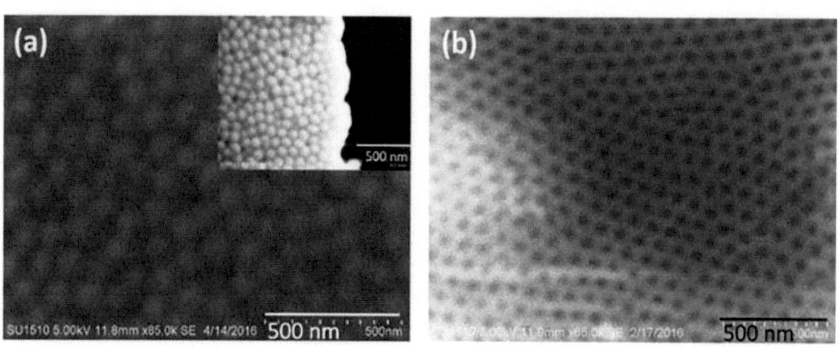

Figure 13. SEM images of bottom side of AAO template: (a) hemispherical cap of barrier layer formed during anodization and (b) complete barrier layer etching in 5% phosphoric acid at 30°C for 30min.

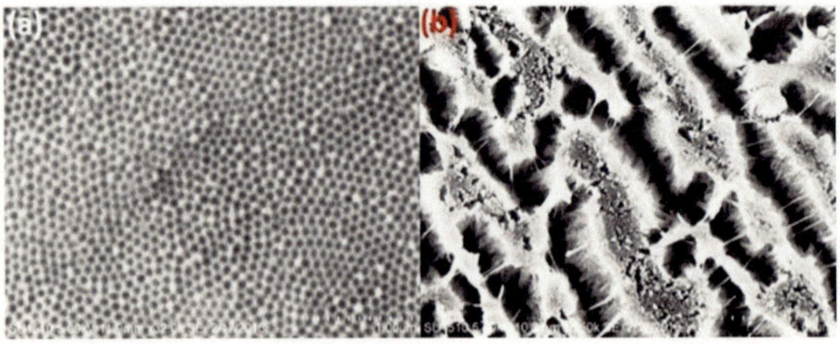

Figure 14. SEM images of bottom and top surfaces after etching in 5% phosphoric acid (H_3PO_4) for 35 min at 30°C. (a) SEM image of bottom surface, which shows all the barrier layers are etched and pores are uniformly open and (b) SEM image of the same sample as in (a), which shows over etching of top surface which looks like pores are bundled together.

surfaces after etching in 5% phosphoric acid (H_3PO_4) for 35 min at 30°C. **Figure 15(a)** shows SEM image of bottom surface, which shows all the barrier layers are etched and pores are uniformly open and **Figure 15(b)** shows SEM image of the same sample as shown in **Figure 13**, which shows over etching of top surface which looks like pores are bundled together.

To check the anodization rate, cross-sectional image of AAO template has been taken at two different times of second anodization, 5 and 250 min, respectively. **Figure 15** shows SEM images of cross-section of homemade templates. **Figure 15(a)** shows ~50-μm long pore which was obtained after 250 min of second anodization, and inset shows smooth and straight pore and; **Figure 15(b)** shows ~ 5-μm long pore which was obtained after 15 min of second anodization. Based on **Figure 15**, anodization rate in the abovementioned condition for the growth of homemade AAO template is 1-μm long in 5 min, which is a faster rate than any other anodization conditions reported before [49] .

As mentioned earlier, removal of the barrier layer from the bottom of the AAO membrane is very challenging; if the etching time is not well controlled in most cases, it results in collapse of the tubular walls and destruction of the pores from the top surface of the template. Results of barrier removal using 1 M H_3PO_4 at 30°C at different times are presented for

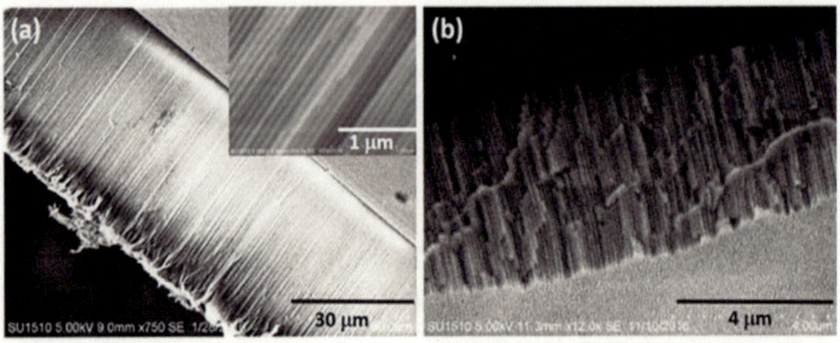

Figure 15. SEM images of cross-section of homemade template: (a) shows ~50-μm long pore which was obtained after 250 min of second anodization, and inset shows smooth and straight pore and (b) shows ~5-μm long pore which was obtained after 15 min of second anodization.

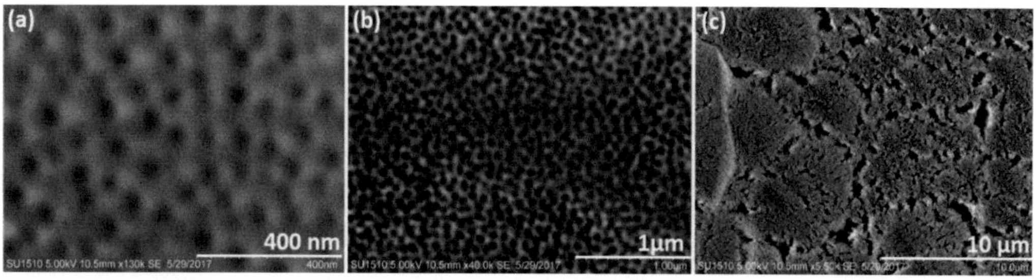

Figure 16. SEM images of top surface of AAO template: (a) shows nonporous hexagonal AAO pores without any etching, (b) after 17 min etching in 1 M phosphoric acid at 30°C and, (c) after 20 min etching in 1 M phosphoric acid at 30°C.

comparison. **Figure 16** shows the SEM images of AAO top surface (a) as-grown and without etching, (b) after 17 min etching in 1 M phosphoric acid at 30°C and (c) after 20 min etching. It is clear from SEM image in **Figure 16(b)** that after 17 min of etching in 1 M phosphoric acid, the AAO template top pores are widened to its maximum capacity, and any increase in etching time will result in merging of the pores. **Figure 16(c)** shows that 20 min of etching is over etching and the pores are merged. After appropriate time etching, barrier layer is thinned and pore diameter is widened. Now, pores can be filled with cobalt by AC electrochemical deposition using a 5% $CoSO_4.7H_2O$ solution stabilized with 2% H_3BO_3 at 20 V AC and 250 Hz [50].

Barrier layer thinning is an alternative technique to remove the barrier layer formed at the end of second anodization process. After thinning of barrier layer, individual pores terminate in the metallic aluminum layer at the bottom of the template. Such membranes grown directly on aluminum foil are suitable for electrodeposition since the bottom aluminum layer works as one electrode.

As discussed in Section 3, nanowire growth can also be carried out in the pores of a track-etched polycarbonate membrane.

5. Semiconducting InSb nanowires grown in polycarbonate template

Following the pioneering work of pore creation in track-etched mica by Possin [51], metallic (Ag) nanowires were grown in 8 nm wide pores by Williams and Giordano [52]. Penner and Martin [53–55] subsequently created pores in polycarbonate membranes by track-etch method. In this work, polycarbonate membranes were purchased from Whatman and had pore lengths of 20 μm and diameter of the order of 200 nm. Following metal deposition on one surface of this template, InSb nanowires were grown in its pores. Unlike the pores in the AAO template, the polycarbonate pores were of uniform diameter and had smooth surfaces. This results in a very smooth surface for the as-grown InSb nanowires.

As-grown InSb nanowires in polycarbonate were first dissolved in dichloromethane and then in chloroform. After InSb nanowires were also grown in track-etched polycarbonate membranes and an SEM image of the as-grown nanowires is shown in **Figure 17**. The biggest

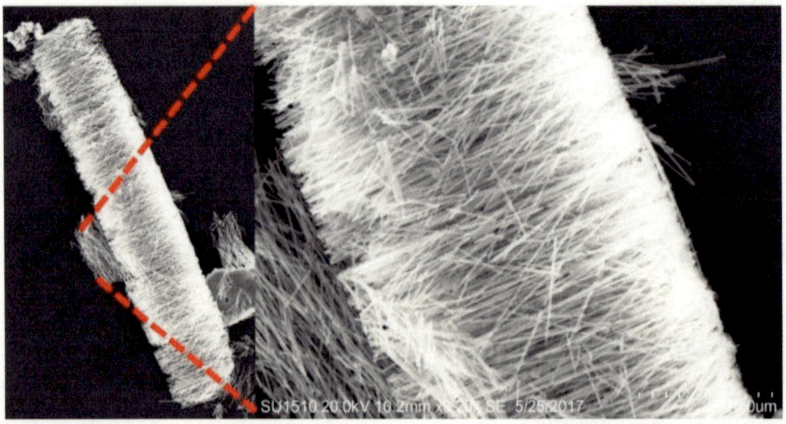

Figure 17. SEM image of a piece of InSb nanowire bundle which was drop-casted on the silicon wafer after dissolving as-grown InSb in polycarbonate template.

challenge in the polycarbonate membrane growth process is the clumping of the nanowires into bundles. This is most likely caused by the residues of the polymer membrane which tends to hold the nanowires together. Dissolution of the membrane is typically done using dichloromethane and chloroform, followed by cleaning in alcohol and DI water, and the nanowires grown by this technique were found to have very smooth surfaces and a length of approximately 20 μm. Electrolyte concentration and cell design were same as used to grow the InSb nanowire in AAO template as discussed in Section 3 except for the pH of solution was maintained at 1.9 instead of 1.7.

Figure 18(a) and **(b)** shows the SEM images of dispersed on silicon substrate at different places and **Figure 18(c)** shows the EDX spectrum of the InSb nanowires and they are In-rich.

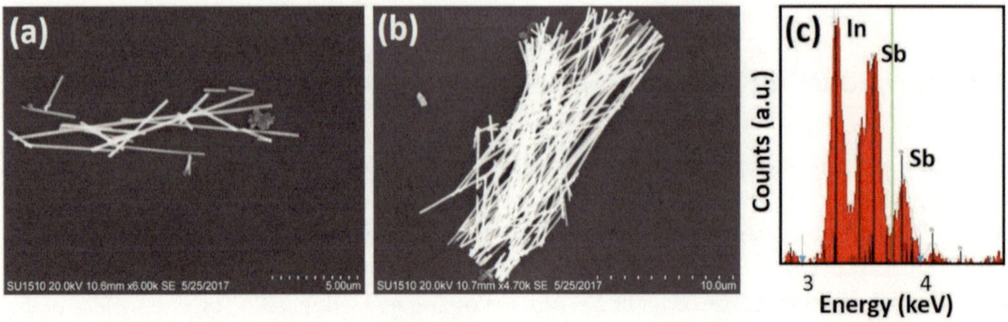

Figure 18. SEM image of InSb nanowire after drop casting on silicon wafer after ultrasonication for 1 min and 1 min oxygen plasma etching: (a) showing broken nanowire due to ultrasonication which resulted in reduced lengths of nanowire from ~20 micron to average 5 micron and (b) showing reduced bundle with approximately 20 μm long InSb nanowires. Average diameter of the nanowires are approximately 150 nm and (c) energy dispersive X-ray (EDX) spectrum of in-rich (54.4 wt% In and 45.6 wt% Sb) electrochemically InSb nanowires grown in polycarbonate template with −1.5 V potential at room temperature and pH = 1.9.

Major challenge is the pore wettability, if this step is not optimized prior to nanowire growth, then even if all other conditions are met, this will result in either no growth or in nonuniform growth in some pores and formation of a mushroom-like crust. To avoid this, it is important that a few basic steps to be performed prior to nanowire growth. Pore-wetting can be achieved by: (i) sonication of nanoporous template in water to remove any air bubbles at the bottom of the pores and (ii) aeration of nanoporous template in vacuum chamber to remove any air bubbles at the bottom of the template. Another challenge is the nonuniform growth which can be avoided by maintaining a temperature gradient between working electrode and mouth of the pore.

6. Conclusions

The growth process of semiconducting nanowires in template pores is a relatively simple process that can be used for synthesizing a high density of well-ordered nanowires in an array. However, the growth process involves several complex basic mechanisms that are affected by cell design, pore wettability, and electrolyte condition. This chapter presents a brief review of work done in this area. It has been experimentally found that commercial AAO template pores have rough tubular surfaces and this results in nanowires with jagged edges. Most commercial AAO templates with narrow pores (of the order of 100 nm) are sometimes tapered and therefore, the metal evaporation and pore-wetting of the pores in these templates are very critical. To circumvent this problem, homemade templates with controlled thickness and pore diameters can be used. However, in this case, the challenge is with the barrier layer formed on the backside of the template. A controlled etching process typically achieves barrier layer removal, where the etching time and etchant concentration as well as temperature are controlled. If the process is not optimized, this will cause the template to tear along the top surface. The tear does not extend over the entire template thickness. The advantage of homemade templates is the pores are relatively uniform and therefore, the nanowires grown in these pores have smooth surfaces. Such smooth nanowires can also be grown in track-etched polycarbonate membranes.

Author details

Abhay Singh* and Usha Philipose

*Address all correspondence to: abhaysingh@my.unt.edu

Department of Physics, University of North Texas, Denton, TX, USA

References

[1] Fan Z, Razavi H, Do J, Moriwaki A, Ergen O, Chueh Y-L, et al. Three-dimensional nanopillar-array photovoltaics on low-cost and flexible substrates. Nature Materials. 2009 Aug;**8**(8):648-653

[2] Banerjee P, Perez I, Henn-Lecordier L, Lee SB, Rubloff GW. Nanotubular metal–insulator–metal capacitor arrays for energy storage. Nature Nanotechnology. 2009 May;**4**(5):292-296

[3] Liu C, Gillette EI, Chen X, Pearse AJ, Kozen AC, Schroeder MA, et al. An all-in-one nanopore battery array. Nature Nanotechnology. 2014 Dec;**9**(12):1031-1039

[4] Tomioka K, Yoshimura M, Fukui T. A III-V nanowire channel on silicon for high-performance vertical transistors. Nature. 2012 Aug 9;**488**(7410):189-192

[5] Luk'yanchuk B, Zheludev NI, Maier SA, Halas NJ, Nordlander P, Giessen H, et al. The Fano resonance in plasmonic nanostructures and metamaterials. Nature Materials. 2010 Sep;**9**(9):707-715

[6] Li KH, Liu X, Wang Q, Zhao S, Mi Z. Ultralow-threshold electrically injected AlGaN nanowire ultraviolet lasers on Si operating at low temperature. Nature Nanotechnology. 2015 Feb;**10**(2):140-144

[7] Kelzenberg MD, Boettcher SW, Petykiewicz JA, Turner-Evans DB, Putnam MC, Warren EL, et al. Enhanced absorption and carrier collection in Si wire arrays for photovoltaic applications. Nature Materials. 2010 Mar;**9**(3):239-244

[8] Pan C, Dong L, Zhu G, Niu S, Yu R, Yang Q, et al. High-resolution electroluminescent imaging of pressure distribution using a piezoelectric nanowire LED array. Nature Photonics. 2013 Sep;**7**(9):752-758

[9] Zhang X, Hao Y, Meng G, Zhang L. Fabrication of highly ordered InSb nanowire arrays by electrodeposition in porous anodic alumina membranes. Journal of the Electrochemical Society. 2005 Oct 1;**152**(10):C664-C668

[10] Kuo C-H, Wu J-M, Lin S-J. Room temperature-synthesized vertically aligned InSb nanowires: Electrical transport and field emission characteristics. Nanoscale Research Letters. 2013;**8**(1):69

[11] Khan MI, Penchev M, Jing X, Wang X, Bozhilov KN, Ozkan M, et al. Electrochemical growth of InSb nanowires and report of a single nanowire field effect transistor. Journal of Nanoelectronics and Optoelectronics. 2008 Jul 1;**3**(2):199-202

[12] Kuo C-H, Wu J-M, Lin S-J, Chang W-C. High sensitivity of middle-wavelength infrared photodetectors based on an individual InSb nanowire. Nanoscale Research Letters. 2013 Dec 1;**8**(1):1-8

[13] Khan MI, Wang X, Jing X, Bozhilov KN, Ozkan CS. Study of a single InSb nanowire fabricated via DC electrodeposition in porous templates. Journal of Nanoscience and Nanotechnology. 2009 Apr;**9**(4):2639-2644

[14] Yang Y, Li L, Huang X, Ye M, Wu Y, Li G. Fabrication of InSb-core/alumina-sheath nanocables. Materials Letters. 2006 Feb;**60**(4):569-571

[15] Vogel AT, de Boor J, Wittemann JV, Mensah SL, Werner P, Schmidt V. Fabrication of high-quality InSb nanowire arrays by chemical beam Epitaxy. Crystal Growth and Design. 2011;**11**(5):1896-1900

[16] Nilsson HA, Caroff P, Thelander C, Larsson M, Wagner JB, Wernersson L-E, et al. Giant, level-dependent g factors in InSb nanowire quantum dots. Nano Letters. 2009 Sep 9;**9**(9): 3151-3156

[17] Vurgaftman I, Meyer JR, Ram-Mohan LR. Band parameters for III–V compound semiconductors and their alloys. Journal of Applied Physics. 2001 Jun 1;**89**(11):5815-5875

[18] Das SR, Delker CJ, Zakharov D, Chen YP, Sands TD, Janes DB. Room temperature device performance of electrodeposited InSb nanowire field effect transistors. Applied Physics Letters. 2011 Jun 13;**98**(24):243504

[19] Solin SA, Thio T, Hines DR, Heremans JJ. Enhanced room-temperature geometric magnetoresistance in inhomogeneous narrow-gap semiconductors. Science. 2000 Sep 1;**289** (5484):1530-1532

[20] Penner RM, Martin CR. Controlling the morphology of electronically conductive polymers. Journal of the Electrochemical Society. 1986 Oct 1;**133**(10):2206-2207

[21] Valizadeh S, George JM, Leisner P, Hultman L. Electrochemical deposition of co nanowire arrays; quantitative consideration of concentration profiles. Electrochimica Acta. 2001 Dec 14;**47**(6):865-874

[22] Eftekhari A. Nanostructured Materials in Electrochemistry. John Wiley & Sons; Weinheim, Germany. 2008. 491 p

[23] Ghahremaninezhad A, Dolati A. Diffusion-controlled growth model for electrodeposited cobalt nanowires in highly ordered aluminum oxide membrane. ECS Transactions. 2010 Oct 8;**28**(17):13-25

[24] Blanco S, Vargas R, Mostany J, Borrás C, Scharifker BR. Modeling the growth of nanowire arrays in porous membrane templates. Journal of the Electrochemical Society. 2014 Jan 1;**161**(8):E3341-E3347

[25] Shin S, Al-Housseiny TT, Kim BS, Cho HH, Stone HA. The race of nanowires: Morphological instabilities and a control strategy. Nano Letters. 2014 Aug 13;**14**(8):4395-4399

[26] Yoo W-C, Lee J-K. Field-dependent growth patterns of metals electroplated in nanoporous alumina membranes. Advanced Materials. 2004 Jul 5;**16**(13):1097-1101

[27] Shin S, Cho HH. Self-formed platform for in situ measurement of electrical transport of individual copper nanowires. Electrochimica Acta. 2014 Jan 20;**117**:120-126

[28] Mebed AM, Abd-Elnaiem AM, Al-Hosiny NM. Electrochemical fabrication of 2D and 3D nickel nanowires using porous anodic alumina templates. Applied Physics A: Materials Science & Processing. 2016 Jun 1;**122**(6):565

[29] Feng Y, Kim K-D, Nemitz CA, Kim P, Pfadler T, Gerigk M, et al. Uniform large-area free-standing silver nanowire arrays on transparent conducting substrates. Journal of the Electrochemical Society. 2016 Jan 1;**163**(8):D447-D452

[30] Khedim MB, Cagnon L, Garagnon C, Serradeil V, Bourgault D. Direct electrical transport measurement on a single thermoelectric nanowire embedded in an alumina template. Physical Chemistry Chemical Physics. 2016 Apr 28;**18**(17):12332-12337

[31] Manzano CV, Bürki G, Pethö L, Michler J, Philippe L. Determining the diffusion mechanism for high aspect ratio ZnO nanowires electrodeposited into anodic aluminum oxide. Journal of Materials Chemistry C. 2017 Feb 16;**5**(7):1706-1713

[32] Singh A, Algarni Z, Philipose U. Template-assisted electrochemical synthesis of p-type InSb nanowires. ECS Journal of Solid State Science and Technology. 2017 Jan 1;**6**(5): N39-N43

[33] Konishi Y, Motoyama M, Matsushima H, Fukunaka Y, Ishii R, Ito Y. Electrodeposition of Cu nanowire arrays with a template. Journal of Electroanalytical Chemistry. 2003 Nov 15;**559**:149-153

[34] APS – APS March Meeting 2010 – Event – Enhanced Raman Scattering from InSb Nanodots; Temperature and Laser-Power Dependent Studies. In: Bulletin of the American Physical Society [Internet]. American Physical Society; Portland, Oregon. 2016 Nov 7. Available from: http://meetings.aps.org/Meeting/MAR10/Session/L9.13

[35] Khan MI, Wang X, Bozhilov KN, Ozkan CS. Templated fabrication of InSb nanowires for Nanoelectronics. Journal of Nanomaterials. 2008 Jul 27;**2008**:e698759

[36] Hnida K, Mech J, Sulka GD. Template-assisted electrodeposition of indium–antimony nanowires – Comparison of electrochemical methods. Applied Surface Science. 2013 Dec 15;**287**:252-256

[37] Kuo C-H,Wu J-M, Lin S-J. Room temperature-synthesized vertically aligned InSb nanowires: Electrical transport and field emission characteristics. Nanoscale Research Letters. 2013;**8**(1):69

[38] Höglund A, Castleton CWM, Göthelid M, Johansson B, Mirbt S. Point defects on the (110) surfaces of InP, InAs, and InSb: A comparison with bulk. Physical Review B. 2006 Aug 29; **74**(7). DOI: https://doi.org/10.1103/PhysRevB.74.075332. https://journals.aps.org/prb/abstract/10.1103/PhysRevB.74.075332

[39] Cui Y, Duan X, Hu J, Lieber CM. Doping and electrical transport in silicon nanowires. The Journal of Physical Chemistry. B. 2000 Jun 1;**104**(22):5213-5216

[40] Wunnicke O. Gate capacitance of back-gated nanowire field-effect transistors. Applied Physics Letters. 2006 Aug 21;**89**(8):083102

[41] Yang Z, Han N, Wang F, Cheung H-Y, Shi X, Yip S, et al. Carbon doping of InSb nanowires for high-performance p-channel field-effect-transistors. Nanoscale. 2013 Sep 27;**5**(20): 9671-9676

[42] Brudnyi VN, Boiko VM, Kamenskaya IV, Kolin NG. Electrical properties and limiting position of the fermi level in InSb irradiated with protons. Semiconductors. 2004;**38**(7): 769-774

[43] Yang YW, Li L, Huang XH, Ye M, YC W, Li GH. Transport properties of InSb nanowire arrays. Applied Physics A: Materials Science & Processing. 2006 Mar 31;**84**(1-2):7-9

[44] Garcia-Vergara SJ, Skeldon P, Thompson GE, Habazaki H. Stress generated porosity in anodic alumina formed in sulphuric acid electrolyte. Corrosion Science. 2007 Oct 1;**49**(10): 3772-3782

[45] Jessensky O, Müller F, Gösele U. Self-organized formation of hexagonal pore structures in anodic alumina. Journal of the Electrochemical Society. 1998 Nov 1;**145**(11):3735-3740

[46] Porous Anodic Aluminum Oxide: Anodization and Templated Synthesis of Functional Nanostructures – Chemical Reviews (ACS Publications) [Internet]. 2017 Aug 23. Available from: http://pubs.acs.org/doi/abs/10.1021/cr500002z

[47] Alam KM, Singh AP, Bodepudi SC, Pramanik S. Fabrication of hexagonally ordered nanopores in anodic alumina: An alternative pretreatment. Surface Science. 2011 Feb;**605** (3-4):441-449

[48] Singh AP, Bodepudi SC, Alam K, Pramanik S. High Density Integration of Carbon Nanotube Spin Valves. In: 2011 International Conference on Nanoscience, Technology and Societal Implications (NSTSI). IEEE; 2011. pp. 1-4. DOI: 10.1109/NSTSI.2011.6111802. http://ieeexplore.ieee.org/document/6111802/

[49] Lee W, Park S-J. Porous anodic aluminum oxide: Anodization and templated synthesis of functional nanostructures. Chemical Reviews. 2014 Aug 13;**114**(15):7487-7556

[50] Jeong S-H, Hwang H-Y, Lee K-H, Jeong Y. Template-based carbon nanotubes and their application to a field emitter. Applied Physics Letters. 2001 Mar 28;**78**(14):2052-2054

[51] Possin GE. A method for forming very small diameter wires. Review of Scientific Instruments. 1970 May 1;**41**(5):772-774

[52] Williams WD, Giordano N. Fabrication of 80 Å metal wires. Review of Scientific Instruments. 1984 Mar 1;**55**(3):410-412

[53] Penner RM, Martin CR. Controlling the morphology of electronically conductive polymers. Journal of the Electrochemical Society. 1986 Oct 1;**133**(10):2206-2207

[54] Cai Z, Martin CR. Electronically conductive polymer fibers with mesoscopic diameters show enhanced electronic conductivities. Journal of the American Chemical Society. 1989 May 1;**111**(11):4138-4139

[55] Schönenberger C, van der Zande BMI, Fokkink LGJ, Henny M, Schmid C, Krüger M, et al. Template synthesis of nanowires in porous polycarbonate membranes: Electrochemistry and morphology. The Journal of Physical Chemistry. B 1997 Jul 1;**101**(28):5497-5505

Chapter 3

Semiconductor Quantum Wells with BenDaniel-Duke Boundary Conditions and Janus Nanorods

Victor Barsan

Additional information is available at the end of the chapter

http://dx.doi.org/10.5772/intechopen.73837

Abstract

The energy levels of bound states of an electron in a quantum well with BenDaniel-Duke boundary condition are studied. Analytic, explicit, simple, and accurate formulae have been obtained for the ground state and the first excited state. In our approach, the exact, transcendental eigenvalues equations were replaced with approximate, tractable, algebraic equations, using algebraic approximations for some trigonometric functions. Our method can be applied to both type I and type II semiconductors and easily extended to quantum dots. The same approach was used for the semi-quantitative analyze of two toy models of Janus nanorods.

Keywords: type I and type II semiconductors, BenDaniel-Duke boundary conditions, Janus nanorods, toy models

1. Introduction

In the last three decades, nanophysics became a domain of increasing interest and intense research, due to the huge number of new effects produced at nanoscale level, in quantum wells (QWs), quantum dots (QDs), Janus nanoparticles, etc. These new effects are fascinating from the perspective of both applied and theoretical physics. The semiconductors provide the largest area of challenging subjects, due to their applications in nanoelectronic devices, multifunctional catalysis, (bio-)chemical sensors, data storage, solar energy conversion, etc.

An attractive aspect of nanophysics is the fact that a quite large number of interesting problems can be approached using quite simple theoretical tools, sometimes at the level of one-particle quantum mechanics. In some cases, the properties of nanostructures like quantum wells, quantum dots, or quantum rods can be explained by just solving the Schrodinger equation with simple potentials. For instance, the basic physical properties of a heterostructure consisting of a thin layer of a semiconductor *A* sandwiched between two somewhat larger semiconductors of identical composition, *B* can be obtained from the study of the movement of a particle with position-dependent mass (PDM) in a finite square well. This particle is, of course, a charge carrier in the semiconductor, and—in our case—will be an electron. As the effective mass of a charge carrier in a semiconductor depends on the charge carrier-lattice interaction, it changes if the lattice composition or the symmetry changes. So, excepting the case of a charge moving in a perfect crystal, the effective mass of an electron or hole is, rigorously speaking, position dependent.

But, simple as the theoretical tools needed for its investigation are, this problem of quantum mechanics involves two important issues: the position-dependent mass (PDM) quantum physics and semiconductor heterostructures. Let us shortly comment on these points.

The roots of the position-dependent effective mass concept are to be found in the pioneering works of Wannier (1937) and Slater (1949) (see Ref. 1 in [1]). Recent papers give explicit methods to obtain explicit solutions of the Schrödinger equation with PDM, for various forms of this dependence and for several classes of potentials [2–4].

However, in practical situations usually encountered in the physics of semiconductor junctions of two materials, *A* and *B*, the simplest and more popular form of position dependence of the effective mass is a step function: the effective mass has a constant value in the material *A* and another, constant value, in the material *B*. In such a case, the most convenient approach for obtaining the wave functions or the envelope functions in a heterostructure—for instance, a quantum well (QW) or quantum dot (QD)—is to solve the Schrödinger equation with BenDaniel-Duke boundary conditions for the wave function [5, 6].

The transition from the complex problem of a real semiconductor (for instance, Kane theory) to the simple problem of a particle moving in a square well with BenDaniel-Duke boundary conditions is indicated, for instance, in Chapter III of Bastard's book [5]. This simple problem provides, however, a realistic description of states near the high-symmetry points in the Brillouin zone of a large class of semiconductors. "It [i.e., 'the simple problem'] often leads to analytical results and leaves the user with the feeling that he can trace back, in a relatively transparent way, the physical origin of the numerical results." ([5], p. 63).

The boundary conditions for the wave functions or envelope functions at interfaces generate the eigenvalue equations for energy; of course, different boundary conditions generate different eigenvalue equations. They are transcendental equations, involving algebraic, trigonometric, hyperbolic, or even more complicated functions. With few exceptions (for instance, the Lambert equation [7]), their solutions, which cannot be expressed as a finite combinations of elementary functions, are not systematically studied.

However, in some situations, quite accurate analytical approximate solutions can be obtained. When a transcendental equation mixes algebraic and trigonometric functions, it might be possible to approximate the trigonometric functions with algebraic expression, and to transform, in this way, the exact transcendental equation into an approximate algebraic one. In its simplest form, for instance, in approximations like $\sin x \simeq x$, for $x \ll 1$, this "algebraization" is largely used. But what is really interesting is to use algebraic approximations of the trigonometric functions valid on their whole domain of definition, as de Alcantara Bonfin and Griffiths proposed in a recent paper [8]; such analytical approximations have been studied and extended by other authors [9].

In this chapter, we shall obtain approximate analytical results for the energy of electronic bound states in quantum wells and in simple models of Janus semiconductor nanorods. As the concept of Janus nanoparticle is less popular than the concept of QWs or QDs, we shall give here some short explanations.

Their name derives from the Roman god Janus: his head had two opposite faces. A Janus nanodot can be a sphere composed of two semispheres of different materials. A Janus nanorod can be a nanorod having the left half and the right half made of different materials. Due to their intrinsic duality, the opposite parts of Janus particles can be functionalized differently [10]. Janus particles with an electron-donor and -acceptor side may be used in photovoltaics. As the Janus nanoparticles have lower symmetry than their homogenous counterparts, their theoretical description is more difficult. In this chapter, we shall propose toy models for semiconductor Janus nanorods.

The structure of this chapter is the following. We shall firstly formulate the basic theory for the quantum mechanical problem of a quantum well, composed of a thin semiconductor sandwiched between two massive ones. This heterojunction can de-modeled by a quantum well (QW), essentially a finite square well, with BenDaniel-Duke boundary conditions. Such a problem was recently discussed by several authors, like Singh et al. [11, 12], who replaced the trigonometric functions entering in the transcendental equations for the bound states energy by the first few terms of their series expansion; in this way, the equations become simple, tractable algebraic ones. Our approach is different, being based on a more sophisticate "algebraization" of trigonometric functions, as proposed by de Alcantara Bonfim and Griffiths [8]. We shall obtain explicit formulas (series expansions) for the ground state energy and for the first excited state, very accurate if the well is not too shallow. Our results can be applied to both type I and type II semiconductors.

In the last part of our chapter, we shall study two toy models for semiconductor Janus nanorods; for the simplest one, we shall obtain analytical expressions for some energy eigenvalues of electronic bound states.

2. Basic theory

We shall solve the Schrodinger equation for an electron moving in a square well, described by the potential:

$$V(x) = \begin{cases} 0, & |x| \leqslant L/2 \\ V_0 > 0, & |x| > L/2 \end{cases} \tag{1}$$

considering that its mass is position dependent. More exactly, the mass inside the well, m_i, and the mass outside the well, m_o, are different:

$$m(x) = \begin{cases} m_i, & |x| \leqslant L/2 \\ m_o, & |x| > L/2 \end{cases} \tag{2}$$

So, the Schrodinger equation for bound states is:

$$H\psi(x) = \left[-\frac{\hbar^2}{2}\frac{d}{dx}\left(\frac{1}{m(x)}\frac{d}{dx}\right) + V(x) \right] \psi_n(x) = E_n \psi_n(x) - \tag{3}$$

Its physically acceptable solutions, that is, the wave functions, have to satisfy two conditions: (1) the continuity of the wave function and (2) the continuity of the probability currents density at the interface. The first one is encountered in all quantum mechanical problems, but the second one is specific to the case of the position-dependent mass [6], defined by the Eq. (2), and takes the form:

$$\frac{1}{m_i}\frac{d\psi_{in}(x < L/2)}{dx}\bigg|_{x \to L/2} = \frac{1}{m_o}\frac{d\psi_{out}(x > L/2)}{dx}\bigg|_{x \to L/2} \tag{4}$$

Eq. (4) is known as the BenDaniel-Duke boundary condition. The notations ψ_{in}, ψ_{out} were used here to make more visible the physical content of this special boundary condition, and will not be maintained in the rest of the chapter.

The nth bound state has a unique energy, E_n, but two wave vectors, one inside the well, $k_{in,n}$, and another one outside, $k_{out,n}$:

$$E_n = \frac{\hbar^2 k_{in,n}^2}{2m}, \quad V_0 - E_n = \frac{\hbar^2 k_{out,n}^2}{2m} \tag{5}$$

Due to the parity of the potential, $V(x) = V(-x)$, the wave functions can be chosen to be symmetric or antisymmetric.

The symmetric wave functions, describing the even states, are:

$$\psi_{2n}(x,\ 0 < x \leqslant L/2) = A_{2n} \cos k_{in,2n} x; \quad \psi_{2n}(x,\ x > L/2) = B_{2n} \exp\left(-k_{out,2n} x\right) \tag{6}$$

$$\psi_{2n}(x < 0) = \psi_{2n}(-x) \tag{7}$$

The ground state wave function is, of course, $\psi_0(x)$. The antisymmetric wave functions, describing the odd states, are:

$$\begin{aligned}\psi_{2n+1}(x,\ \ 0 < x \leqslant L/2) &= A_{2n+1} \sin k_{in,\,2n+1} x \\ \psi_{2n+1}(x,\ \ x > L/2) &= B_{2n+1,s} \exp\left(-k_{out,\,2n+1} x\right)\end{aligned} \tag{8}$$

$$\psi_{2n+1}(x < 0) = -\psi_{2n+1}(-x) \tag{9}$$

The continuity of these functions in $x = L/2$ gives:

$$B_{2n} = A_{2n} \cos \frac{k_{in,\,2n} L}{2} \exp\left(\frac{k_{out,\,2n} L}{2}\right) \tag{10}$$

$$B_{2n+1} = A_{2n+1} \sin \frac{k_{in,\,2n+1} L}{2} \exp\left(\frac{k_{out,\,2n+1} L}{2}\right) \tag{11}$$

So, the wave function outside the well is:

$$\psi_{2n}(x > L/2) = A_{2n} \cos \frac{k_{in,\,2n} L}{2} \exp\left(-k_{out,\,2n}\left(x - \frac{L}{2}\right)\right) \tag{12}$$

$$\psi_{2n+1}(x > L/2) = A_{2n+1} \sin \frac{k_{in,\,2n+1} L}{2} \exp\left(-k_{out,\,2n+1}\left(x - \frac{L}{2}\right)\right) \tag{13}$$

The wave functions are normalized if:

$$\frac{1}{A_{2n}^2} = \frac{L}{2}\left(1 + \frac{\sin k_{in,\,2n} L}{k_{in,\,2n} L} + \frac{1 + \cos k_{in,\,2n} L}{k_{out,\,2n} L}\right) \tag{14}$$

$$\frac{1}{A_{2n+1}^2} = \frac{L}{2}\left(1 - \frac{\sin k_{in,\,2n+1} L}{k_{in,\,2n+1} L} + \frac{1 - \cos k_{in,\,2n+1} L}{k_{out,\,2n+1} L}\right) \tag{15}$$

These results generalize the formula (24) in [11] and the Eqs. (25.3e, o) in [13].

It is convenient to use the potential strength P (introduced by Pitkanen [14], who actually used α, instead of P)

$$P = \sqrt{\frac{L^2}{2\hbar^2} m_i V} \tag{16}$$

and to define also ε_n, β, and X as:

$$\varepsilon_n = \frac{E_n}{V} \tag{17}$$

$$\beta = \frac{m_i}{m_0} \tag{18}$$

$$\Phi_n = k_{in,\,n} \frac{L}{2} \tag{19}$$

P, ε_n, β, and Φ_n are dimensionless quantities; Φ_n will be sometimes called dimensionless wave vector.

It is easy to see that:

$$k_{in,n}\frac{L}{2} = P\sqrt{\varepsilon_n} \tag{20}$$

$$k_{out,n}\frac{L}{2} = P\sqrt{\frac{1-\varepsilon_n}{\beta}} \tag{21}$$

$$k_{in,n}^2 + \beta k_{out,n}^2 = \frac{1}{(pL/2)^2} \tag{22}$$

Let us mention that, if the mass is position-independent, that is, if $m_i = m_o$, the eigenvalue equations are (see for instance [5], p. 3, Eqs. (15) and (16)):

$$\tan\frac{k_{in,2n}L}{2} = \frac{k_{out,2n}}{k_{in,2n}}, \quad even \quad states \tag{23}$$

$$\tan\frac{k_{in,2n+1}L}{2} = -\frac{k_{in,2n+1}}{k_{out,2n+1}}, \quad odd\, states \tag{24}$$

If the mass is position dependent, according to (2), the eigenvalue equations obtained from the Schrodinger equations, using BenDaniel-Duke boundary conditions have the form:

$$\tan\frac{k_{in,2n}L}{2} = \frac{m_i}{m_o}\frac{k_{out,2n}}{k_{in,2n}} == \beta\frac{k_{out,2n}}{k_{in,2n}}, \quad even \quad states \tag{25}$$

$$\tan\frac{k_{in,2n+1}L}{2} = -\frac{k_{in,2n+1}}{k_{out,2n+1}}, \quad odd\, states \tag{26}$$

We shall consider that both m_i, m_o are positive; this corresponds to type I semiconductors. So, with $\beta > 0$, with $k_{in}L/2$ replaced by Φ_{2n} for even states and by Φ_{2n+1} for odd states, we can put the Eqs. (25) and (26) in a more convenient form:

$$\Phi_{2n}\tan\Phi_{2n} = \frac{\sqrt{\beta}}{p}\sqrt{1-p^2\Phi_{2n}^2}, \quad n = 0,\ 1, \ldots even \quad states \tag{27}$$

$$\Phi_{2n+1}\cot\Phi_{2n+1} = -\frac{\sqrt{\beta}}{p}\sqrt{1-p^2\Phi_{2n+1}^2}, \quad n = 0,\ 1, \ldots odd \quad states \tag{28}$$

or, equivalently:

$$\frac{\cos\Phi_{2n}}{\Phi_{2n}} = (-1)^n\frac{p}{\sqrt{\beta+(1-\beta)p^2\Phi_{2n}^2}}, \quad n = 0,\ 1,\ \ldots \quad even \quad states \tag{29}$$

http://dx.doi.org/10.5772/intechopen.73837

$$\frac{\sin\Phi_{2n+1}}{\Phi_{2n+1}} = (-1)^n \frac{p}{\sqrt{\beta + (1-\beta)p^2\Phi_{2n+1}^2}}, \quad n = 0,\ 1,\ \ldots \ odd\ states \tag{30}$$

For $\beta = 1$, they take the form of the well-known equations for the energy eigenvalues of the finite square well. Approximate analytical solutions of these equations were obtained for deep wells ($p \ll 1$) [15] and in the general case [8, 9, 16, 17].

If $0 < \beta < 1$ $(\beta > 1)$, the rhs of Eqs. (29) and (30) is a monotonically decreasing (increasing) function of Φ; in both cases, the roots of these equations can be obtained using the same approach.

In this chapter, we shall obtain precise analytical approximations for the energy of the first two states, that is, for the ground state and for the first excited state, considering the cases $\beta < 1$ and $\beta > 1$ separately. For moderate and deep wells, the formulae are both simple and accurate. In the limit $\beta \to 1$, we shall obtain the result of de Alcantara Bonfim and Griffiths, Eq. (17) of [8].

3. Approximate analytical solutions for eigenvalue equations

3.1. The first even state (the ground state)

According to Eq. (29), the dimensionless momentum of the first even state, which is also the ground state, is the smallest positive root of the equation:

$$\frac{\cos\Phi_0}{\Phi_0} = \frac{p}{\sqrt{\beta + (1-\beta)p^2\Phi_0^2}}, \quad 0 < x < \frac{\pi}{2} \tag{31}$$

We shall discuss separately the cases $\beta > 1$ and $\beta < 1$.

3.1.1. The case $\beta > 1$

It is useful to introduce the new parameters $\gamma_>,\ g_>,\ A_>^2$:

$$\gamma_> = \beta - 1,\ g_> = \frac{1}{\gamma_>},\ A_>^2 = \frac{P^2\beta}{\beta - 1} = P^2\beta g_> \tag{32}$$

because the eigenvalue equation can be written in a simpler form:

$$\frac{\cos\Phi_0}{\Phi_0} = \frac{1}{\sqrt{\gamma_>}\sqrt{A_>^2 - \Phi_0^2}}, \quad 0 < \Phi_0 < \frac{\pi}{2} \tag{33}$$

In the most physically interesting cases, P is quite large (the wells are quite deep), and according to (32), $A_>$ is even larger, so it is more convenient to use A instead of P as "large parameter".

We shall replace the exact, transcendental Eq. (31) with an approximate, algebraic equation, using one of the formulae proposed in [8] for $\cos x$, namely:

$$\cos x \simeq f(x,c) = \frac{1-\left(\frac{2x}{\pi}\right)^2}{\sqrt{1+cx^2}} \tag{34}$$

The precision of this approximation on various subintervals of $(0,1)$ depends on the exact value of c, with $0.18 \lesssim c \lesssim 0.23$; in our numerical evaluation, we shall use the value $c = 0.22$. For a detailed discussion on this issue, see [18].

The algebraic approximation of the eigenvalue equation, we get with (34) is:

$$\frac{1}{\Phi_0}\frac{1-\left(\frac{2\Phi_0}{\pi}\right)^2}{\sqrt{1+c\Phi_0^2}} = \frac{1}{\sqrt{\gamma_>}\sqrt{A_>^2-\Phi_0^2}} \tag{35}$$

with

$$\Phi_0^2 = z \tag{36}$$

(35) can be written as:

$$z^3 + \left(\frac{1}{16}\pi^4 cg_> - A_>^2 - \frac{1}{2}\pi^2\right)z^2 + \frac{\pi^2}{2}\left(\frac{\pi^2}{8}g_> + A_>^2 + \frac{1}{8}\pi^2\right)z - \frac{1}{16}\pi^4 A_>^2 = 0 \tag{37}$$

Following the approach outlined in [19] and applied to this problem in [18], introducing the notation:

$$C = \frac{\pi^2}{2}c \tag{38}$$

and considering that the well is not too shallow:

$$A_>^2 \gg 1$$

we obtain for the physically interesting root the expression:

$$\begin{aligned} z(\beta>1) &= \frac{\pi^2}{4} - \frac{\pi^3}{8}\sqrt{\left(1+\frac{C}{2}\right)}\left(\sqrt{g_>}\alpha_>\right) + \frac{\pi^4}{32}(1+C)\left(\sqrt{g_>}\alpha_>\right)^2 \\ &+ \frac{\pi^5}{32}\left(1+\frac{Cg_>}{2}\right)\sqrt{\left(1+\frac{C}{2}\right)}\sqrt{g_>}\alpha_>^3 + \frac{\pi^6}{128}g_>(1+C)\left(1+\frac{Cg_>}{2}\right)\alpha_>^4 + \dots \end{aligned} \tag{39}$$

If the depth of the well increases indefinitely, $\alpha_> \to 0$ and $z_1 \to \pi^2/4$, $\Phi_0 \to \pi/2$, as requested. Indeed, in a finite well, the energy of a bound state is smaller than the corresponding energy in an infinite one, so the first term in $\sqrt{g_>}\alpha_>$ in the previous formula is negative.

http://dx.doi.org/10.5772/intechopen.73837

It is useful to write (39) in terms of more physical parameters, p and β. In order to do this, let us notice that:

$$g_{>}\alpha_{>}^{2}=\frac{p^{2}}{\beta},\quad \alpha_{>}^{2}=\frac{\beta-1}{\beta}p^{2} \tag{40}$$

so Eq. (39) takes the form:

$$\begin{aligned} z(\beta>1)&=\frac{\pi^{2}}{4}-\frac{\pi^{3}}{8}\sqrt{\left(1+\frac{C}{2}\right)}\frac{p}{\beta^{1/2}}+\frac{\pi^{4}}{32}(1+C)\frac{p^{2}}{\beta}\\ &+\frac{\pi^{5}}{32}\sqrt{1+\frac{C}{2}}\left(\beta-1+\frac{C}{2}\right)\frac{p^{3}}{\beta^{3/2}}+\frac{\pi^{6}}{128}\left(\beta-1+\frac{C}{2}\right)(1+C)\frac{p^{4}}{\beta^{2}}+\dots \end{aligned} \tag{41}$$

and:

$$z(\beta=1)=\frac{\pi^{2}}{4}-\frac{\pi^{3}}{8}\sqrt{\left(1+\frac{C}{2}\right)}p+\frac{\pi^{4}}{32}(1+C)p^{2}+\frac{\pi^{5}}{32}\sqrt{1+\frac{C}{2}}\frac{C}{2}p^{3}+\frac{\pi^{6}}{128}\frac{C}{2}(1+C)p^{4}+\dots \tag{42}$$

It is a simple exercise to check that the first three terms of the previous formula coincides with the first three terms of the power series given by Eq. (17) of [8].

If the parameter $A_{>}$ cannot be considered "large," the exact expression of the root can be obtained using the standard approach [19]; they are elementary, but cumbersome, and will be not given here; the interested reader can find them in [18] .

3.1.2. The case $\beta<1$

If $\beta<1$, the eigenvalue equation for the dimensionless wave vector is:

$$\frac{\cos\Phi_{0}}{\Phi_{0}}=\frac{1}{\sqrt{\gamma_{<}}\sqrt{A_{<}^{2}+\Phi_{0}^{2}}},\quad 0<\Phi_{0}<\frac{\pi}{2} \tag{43}$$

with the following definitions for the parameters:

$$\gamma_{<}=1-\beta,\ \frac{1}{\gamma_{<}}=g_{<},\ A_{<}^{2}=\frac{P^{2}\beta}{1-\beta}=P^{2}\beta\gamma_{<} \tag{44}$$

Using the de Alcantara Bonfim-Griffiths algebraization for $\cos x$ (34) [8], it gives an algebraic equation, which becomes, with the same substitution

$$\Phi_{0}^{2}=z \tag{45}$$

a cubic equation:

$$z^3+\left(A_<^2-\frac{1}{2}\pi^2-\frac{1}{16}\pi^4 cg_<\right)z^2+\frac{\pi^2}{2}\left(\frac{1}{8}\pi^2-A_<^2-\frac{\pi^2}{8}g_<\right)z+\frac{1}{16}\pi^4A_<^2=0, \qquad \beta<1 \quad (46)$$

Following the same steps as in the previous case, we find that the parameters $g_>$, $\alpha_>$ enter into the various expressions needed for obtaining the cubic roots only through the monoms $g_>\alpha_>^2$, $\alpha_>^2$ at various powers, and the roots of Eq. (46) can be obtained from the root (39) making the substitution:

$$g_> \to -g_<, \ \alpha_>^2 \to -\alpha_<^2 \qquad (47)$$

in Eq. (39). The final result, $z(\beta<1)$, expressed in terms of p and β, has exactly the form (41).

3.2. The first odd state

3.2.1. The case $\beta>1$

The exact eigenvalue equation for the first odd state, which is also the first excited state, can be written as:

$$\frac{\sin\Phi_1}{\Phi_1}=\frac{1}{\sqrt{\gamma_>}\sqrt{A_>^2-\Phi_1^2}}, \quad \frac{\pi}{2}<\Phi_1<\pi \qquad (48)$$

As the shape of the function $\sin x/x$ on the interval $[0,\pi]$ is quite similar with the shape of $\cos x$ on the interval $[0,\pi/2]$, we can try an algebraization for $\sin x/x$ similar to that proposed by de Alcantara Bonfim and Griffiths for $\cos x$:

$$\frac{\sin\Phi_1}{\Phi_1}\simeq\frac{1-(\Phi_1/\pi)^2}{\sqrt{1+a\Phi_1^2}}, \quad 0<\Phi_1<\pi, \ a\simeq 0.2 \qquad (49)$$

A detailed discussion of the precision of this approximation is given in [18] (see Fig. 3 and Eq. (88)). Following, exactly the same steps as in the case of the ground state, we find that

$$z(\beta)=\pi^2-\pi^2\sqrt{(1+\pi^2a)}\frac{p}{\sqrt{\beta}}+\frac{\pi^4}{2}a\frac{p^2}{\beta}+\pi^4\sqrt{(1+\pi^2a)}(\beta-1)\left(\frac{p}{\sqrt{\beta}}\right)^3+\frac{\pi^6}{2}a(\beta-1)\frac{p^4}{\beta^2}+\dots \qquad (50)$$

For $\beta<1$, the expression of the root, in terms of β and p is identical with (50), written in terms of $g_<$, $g_>$, $\alpha_<$, $\alpha_>$, the formulae are different, see Eqs. (99) and (103) in [18].

For both cases—$\beta\lessgtr 1$—in the limit of an infinitely deep root, $z(\beta\lessgtr 1, \ \alpha_<=0)=\pi^2$, $\Phi_1(\beta\lessgtr 1, \alpha_<=0)=\pi$, as requested, and the first correction to this value is negative.

The relative errors of the formulas (39) and (50), with respect to the exact roots of the corresponding algebraic equations, are very small—of about $10^{-4}\dots10^{-6}$ for physically interesting values of the parameters $p, \ \beta, \ a, \ c$. In other words, the main contribution to the errors of our results is given by the approximation of trigonometric functions with algebraic ones, not by the approximation of the exact formulae of the roots of cubic equations with the low order terms of their series expansions.

As already mentioned, one of the physical motivations of the calculation of the energy of bound states in heterostructures is to explain their photoluminescence properties. In several cases (see for instance [20]), the authors use Barker's formula for the energy levels in a square well [15]. Much more precise analytical expressions for these energy are available in the literature [8, 9], for the case of constant mass; in this paper, we propose similar formulas, considering the case of position-dependent mass.

3.3. Higher-order states

In the previous subsections, we analyzed the ground state $(n = 0)$ and the first excited state $(n = 1)$ of a square well, with BenDaniel-Duke boundary conditions. For $n \geqslant 2$, the de Alcantara Bonfim formula (34) can be extended to larger arguments:

$$\cos \Phi \simeq \frac{1 - 4(\Phi - 2n\pi)^2/\pi^2}{\sqrt{1 + c(\Phi - 2n\pi)^2}}, \quad 2n\pi < \Phi < 2n\pi + \frac{\pi}{2} \tag{51}$$

but the eigenvalue equation, obtained in this way, is a sextic equation (which cannot be reduced to a cubic equation in Φ^2), so it cannot be solved. We meet similar difficulties if we try to use in the eigenvalue equation the algebraization of tan (see later on, Eq. (73) and (74) of the present paper). Even the "parabolic approximation" for $(\cos x/x)^2$ or $(\sin x/x)^2$, in the sense used in [16], gives a quartic equation for the dimensionless wave vector. Its roots are given by complicated, but still elementary formulas.

3.4. Graphical illustration of our main results

In order to illustrate graphically some of our results, let us notice that, using Eqs. (17)–(22), we can write the following relations for the energy:

$$\frac{m_i L^2}{2\hbar^2} E_n = \Phi_n^2 = z_n(p) \tag{52}$$

where z_n is the root of the cubic equations obtained after the algebraization of the transcendental eigenvalue equations for the ground state $(n = 0)$ and for the first excited state $(n = 1)$. According to the Eqs. (39) and (50), for a deep well, the root z can be approximated with a quartic polynomial in p, the inverse of the potential strength P. Let us mention that, if we replace in the definition of P, Eq. (16), m_i with the free electron mass, we choose the length of the well $L = 10$ nm and we express the potential V_0 in electron volts, we get:

$$P = 25.616\sqrt{V_0}, \ p = 3.9 \times 10^{-2} \frac{1}{\sqrt{V_0}} \tag{53}$$

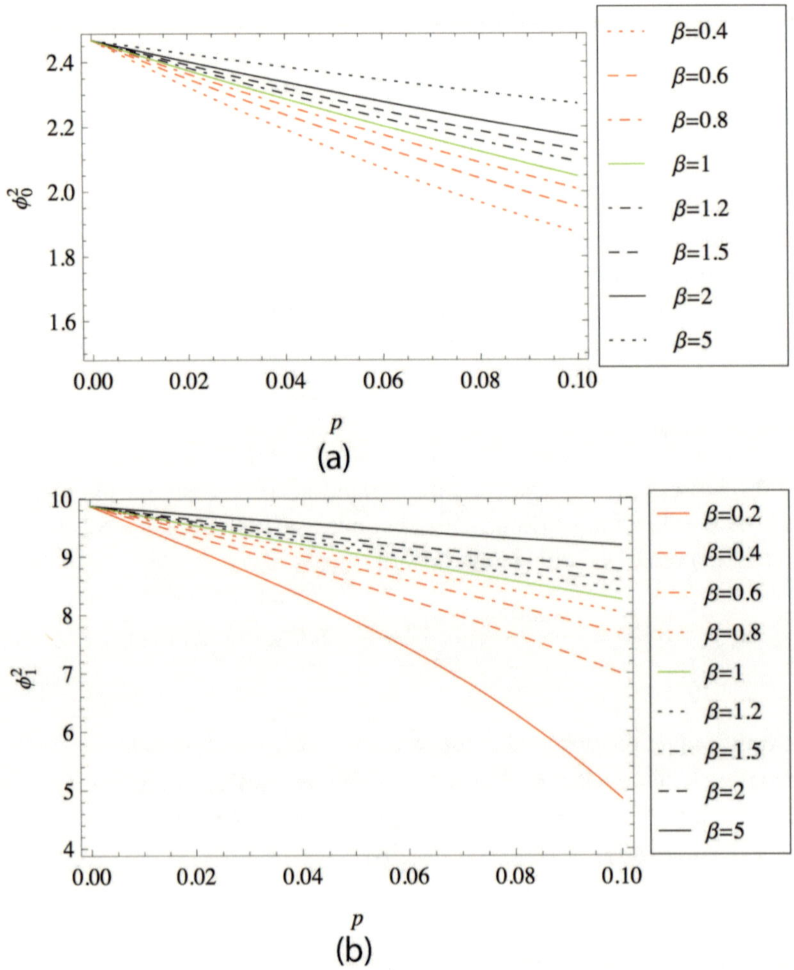

Figure 1. The plot of Φ_0^2 (a) and Φ_1^2 (b), which are proportional to the energies E_0, E_1, as functions of the inverse potential strength p.

We shall plot our main results, that is, the series expansions of the dimensionless wave vectors, Φ_0^2 and Φ_1^2, as functions of p, on the range $0 < p < 0.1$, when the condition of convergence is satisfactorily fulfilled (**Figure 1**). The energy is a monotonically increasing function of β; its values, for $\beta = 1$, are obtained from Eqs. (42) and (50).

3.5. Applications to other nanostructures

Our calculations can be easily applied to type II semiconductors heterostructures, when one of the effective mass of the charge carrier is negative: $m_i m_o < 0$ ([5], chapter 3, Eqs. (35) and (36)); a detailed description of such heterointerfaces can be found for instance in [5], p. 66. So, instead of (25) and (26), the eigenvalue equations take the form:

$$\tan\frac{k_{in,2n}L}{2} = -\frac{m_i}{|m_o|}\frac{k_{out,2n}}{k_{in,2n}} = -|\beta|\frac{k_{out,2n}}{k_{in,2n}}, \quad even \quad states \tag{54}$$

$$\tan\frac{k_{in,2n+1}L}{2} = \frac{|m_o|}{m_i}\frac{k_{in,2n+1}}{k_{out,2n+1}} = \frac{1}{|\beta|}|\beta|, \quad odd\, states \tag{55}$$

and can be solved following exactly the same approach.

As already mentioned, the wave function in the Schrodinger Eq. (3) can be interpreted as an envelope function. This approximation works well when the materials constituting the heterostructures are perfectly lattice-matched and they crystallize in the same crystallographic structure (in the most cases, the zinc blend structure). Its application is restricted to the vicinity of the high-symmetry points in the host's Brillouin zone $(\Gamma, \ X, \ L)$. Actually, most of the heterostructures' energy levels relevant to actual devices are relatively closed to a symmetry

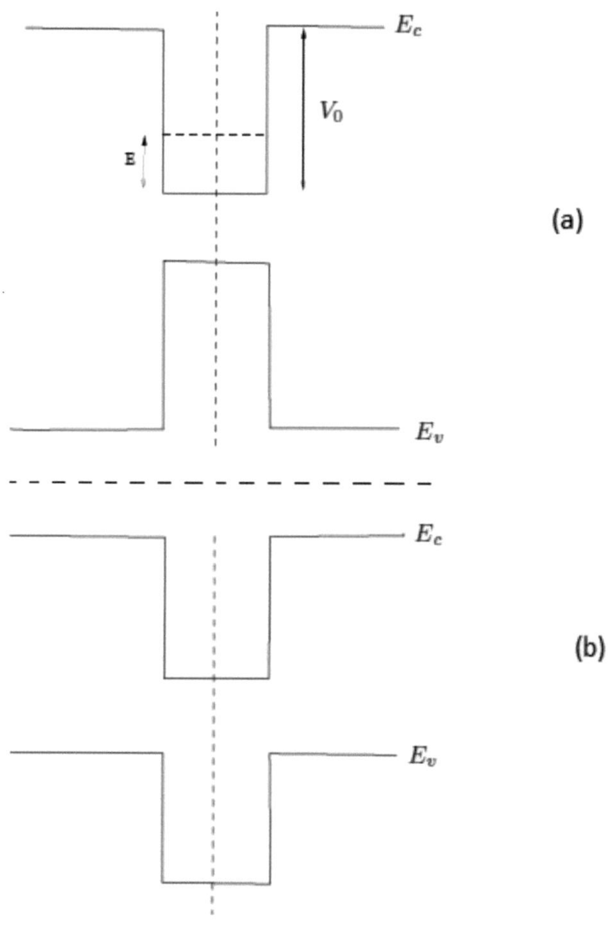

Figure 2. Schematic representation of the conduction band E_c and of the valence band E_v for type I (a) and type (II) (b) semiconductors.

point in the host's Brillouin zone. A popular example is given by the lowest conduction states of GaAs-GaAlAs heterostructures with GaAs layer (typically, its thickness is about 100 Å or larger). A detailed description of the cases in which the envelope function model is successful is given in [5], p. 66 (See **Figure 2**).

As there are some similarities between QWs and QDs, our results are also relevant for these devices. The simplest remark is that the eigenvalues equations for the first odd state in a QW are identical to that corresponding to the $l = 0$ state in a QD (see for instance [13], problem 63). Also, the eigenvalue equations for the wave vectors of the energy levels for a finite barrier rectangular shaped QD, Eq. (36) in [21], are quite similar to ours—(29) and (30), but somewhat more complicated. The ground state energy of electrons and holes in a core/shell QD is given by the Eq. (21) of [22], an equation similar to ours, just mentioned previously. Such results are important, *inter alia,* for the interpretation of photoluminescence spectra and photon harvesting of QDs.

4. The infinite square well with two semiconductor slabs

4.1. The symmetric case

Let us consider an infinite 1D square well, delimited by two rigid walls situated in $-L/2$, respectively $L/2$, containing two semiconductor slabs, of equal width, but of different materials. It is a toy model for a Janus nanorod, composed of two different semiconductors, with large work functions. We preferred to choose this particular case (equal width), in order to avoid too cumbersome mathematical calculations. The electron effective mass is position dependent, like in (2):

$$m(x) = \begin{cases} m_1, & -L/2 < x < 0 \\ m_2, & 0 < x < L/2 \end{cases} \tag{56}$$

with:

$$m_2 = \beta m_1 \tag{57}$$

We want to investigate how the energies of the electronic bound states will be affected, compared to the situation when in the infinite well there is only one slab, with effective electron mass m_1 or m_2. As

$$E = \frac{\hbar^2 k_1^2}{2m_1} = \frac{\hbar^2 k_2^2}{2m_2} \tag{58}$$

we have, with (57):

$$k_2 = \sqrt{\beta} k, \;\; k = k_1 \tag{59}$$

The electronic wave function is obtained solving the Schrodinger equation, as in the case of a finite well, studied in Section 2:

$$\psi(x) = \begin{cases} A_1 \sin\left(k_1 x + \varphi_1\right), & -L/2 < x < 0 \\ A_2 \sin\left(\sqrt{\beta} k_1 x + \varphi_2\right), & -L/2 < x < 0 \end{cases} \tag{60}$$

The boundary conditions for the wave function give:

$$\sin\left(-\frac{kL}{2} + \varphi_1\right) = 0 \tag{61}$$

$$\sin\left(\sqrt{\beta}\frac{kL}{2} + \varphi_2\right) = 0 \tag{62}$$

and the continuity in the origin:

$$A_1 \sin\varphi_1 = A_2 \sin\varphi_2 \tag{63}$$

The BenDaniel-Duke boundary condition means:

$$\frac{1}{m_1}\psi'(0_-) = \frac{1}{m_2}\psi'(0_+) \tag{64}$$

or:

$$\sqrt{\beta} A_1 \cos\varphi_1 = A_2 \cos\varphi_2 \tag{65}$$

Together with the orthonormality condition for the wave function, Eqs. (61)–(63) and (65) form a system of five equations for five quantities, $k,\ \varphi_1,\ \varphi_2,\ A_1,\ A_2$. As the amplitudes are not of primary interest, we can combine (64) and (65) to obtain:

$$\frac{1}{\sqrt{\beta}} \tan\varphi_1 = \tan\varphi_2 \tag{66}$$

Replacing in (66), the values of $\varphi_1,\ \varphi_2$ obtained from (61) and (62):

$$\varphi_1 = n_1\pi + \frac{kL}{2},\ \varphi_2 = n_2\pi - \sqrt{\beta}\frac{kL}{2} \tag{67}$$

we get:

$$\tan\frac{kL}{2} = -\sqrt{\beta}\tan\sqrt{\beta}\frac{kL}{2} \tag{68}$$

With

$$\frac{kL}{2} = K \tag{69}$$

it can be written as:

$$\tan K + \sqrt{\beta} \tan \sqrt{\beta} K = 0 \tag{70}$$

If $\sqrt{\beta} = 1$, (67) gives:

$$2 \tan \frac{kL}{2} = n\pi \rightarrow k_n = \frac{n\pi}{L} \tag{71}$$

$$k_1 = \frac{\pi}{L} \rightarrow K_1 = \frac{\pi}{2} \tag{72}$$

so the solutions corresponding to the infinite well with an homogenous medium inside the walls.

Eq. (70) is a transcendental one, and its solutions cannot be expressed as a finite combination of elementary functions. A quite popular analytical approximation for the tangent function has been proposed by de Alcantara-Bonfim [8] and generalized by the present author [9]:

$$\tan x \simeq \frac{0.45\pi(x - n\pi)}{2x - (2n - 1)\pi}, \quad \left(n - \frac{1}{2}\right)\pi < x < n\pi \tag{73}$$

$$\tan x \simeq \frac{0.45\pi(x - n\pi)}{(2n + 1)\pi - 2x}, \quad n\pi < x < \left(n + \frac{1}{2}\right)\pi \tag{74}$$

In order to see how this approximation works, let us consider the first two roots of Eq. (68), if $\beta \lesssim 1$. For $\beta = 0.9$, we obtain (for instance, using the command FindRoot in Mathematica) $K_{1,exact} = 1.65804$ and $K_{2,exact} = 3.29797$, close to $\pi/2$, respectively π, that is, to the values corresponding to $\beta = 0$. We shall discuss the case of the second root of Eq. (70). As $K_{2,exact} = 3.29797 \in \left(\pi, \frac{3}{2}\pi\right)$ and $K_{2,exact}\sqrt{\beta} = 2.9682 \in \left(\frac{\pi}{2}, \pi\right)$, the two tangent functions appearing in Eq. (70) will be approximated by the two variants of Eqs. (73) and (74), the result being the following:

$$\frac{K - \pi}{3\pi - 2K} + \frac{\sqrt{\beta}(\sqrt{\beta}K - \pi)}{2\sqrt{\beta}K - \pi} = 0 \tag{75}$$

So, K can be obtained as a root of a second order equation, namely:

$$K(\beta) = \frac{\pi}{4} \frac{-3\beta + 1 + \sqrt{9\beta^2 - 24\beta^{\frac{3}{2}} + 26\beta - 8\beta^{1/2} + 1}}{\beta^{1/2}(1 - \beta^{1/2})} \tag{76}$$

We find that $K(\beta = 0.9) = K_{2,approx} = 3.2987$, so quite close to the exact value, the error being:

$$\frac{K_{2,exact} - K_{2,approx}}{K_{2,exact}} = \frac{3.29797 - 3.2987}{3.29797} = -2.2135 \times 10^{-4} \tag{77}$$

However, due to the rapid variation of the tangent functions near its singularities, this approximation method must be used with utmost care, as it can easily give unacceptable results (this is the case of the first root, for $\beta = 0.9$).

4.2. The asymmetric case

Let us consider now the case of a rectangular infinite asymmetric well, with the potential:

$$V(x) = \begin{cases} \infty, & x < -L/2 \\ 0, & -L/2 < x < 0 \\ V_0, & 0 < x < L/2 \\ \infty, & x > L/2 \end{cases} \tag{78}$$

with $V_0 > 0$, containing, as in the previous example, two semiconductor slabs. It is also a toy model of a Janus nanorod, somewhat more realistic than that discussed in Section 4.1. We also chose a particular geometry (the same width for each slab) to avoid irrelevant mathematical complications. For an electronic bound state of energy $E > V_0$, the wave vectors (and the electronic effective masses) are different in different slabs, according to the relations:

$$E = \frac{\hbar^2 k_1^2}{2m_1}; \; E - V_0 = \frac{\hbar^2 k_2^2}{2m_2}, \; m_2 = \beta m_1 \tag{79}$$

Defining the wave vector k_0 by:

$$V_0 = \frac{\hbar^2 k_0^2}{2m_1} \tag{80}$$

noticing that:

$$k_2^2 = \beta \left(k_1^2 - k_2^2 \right) \tag{81}$$

and following exactly the same steps as in the symmetric case, we obtain the following eigenvalue equation:

$$\frac{1}{K_1} \tan K_1 + \frac{\sqrt{\beta}}{\sqrt{K_1^2 - K_0^2}} \tan \sqrt{\beta \left(K_1^2 - K_0^2 \right)} = 0 \tag{82}$$

with the notations:

$$K_1 = \frac{k_1 L}{2}, \; K_2 = \frac{k_2 L}{2}, \; K_0 = \frac{k_0 L}{2} \tag{83}$$

If $0 < E < V_0$, k_2 (and, evidently, K_2) become imaginary, and $\tan K_2 \to -i \tanh K_2$. The Eq. (82) and its hyperbolic counterpart are much complicated than (68); even if there are some methods of obtaining approximate analytical solutions, they will be not discussed here. The case of a finite asymmetric well, with two different semiconductor slices, can be studied following exactly the same approach, but now the complications are even more serious, as the wave function extends outside the wells.

5. Conclusions

In this chapter, we obtained approximate analytical solutions for the eigenvalue equation of the first two bound states in a semiconductor quantum well, in a particular case of position-dependent mass of the charge carrier—in fact, the simplest one, corresponding to BenDaniel-Duke boundary conditions. This position dependence can be characterized by β, the ratio of the mass inside, to the mass outside the well. Actually, we obtained quite simple expressions for the dimensionless wave vector, in terms of the potential strength and of β. Even if we solved this problem in terms of one-particle quantum mechanics, obtaining the wave function and the eigenvalues of the bound states, our results can be directly applied in the theory of envelope functions in the conduction band at heterointerfaces. Our approach is based on the "algebraization" of trigonometric functions present in the transcendental eigenvalue equations; in this way, they are transformed in tractable algebraic equations.

We also proposed two models for a semiconductor Janus nanorod—a system, which was not yet treated analytically.

Our results can be easily extended to more realistic (e.g., linear) position dependence of the mass carrier and to other nanosystems. For instance, the eigenvalue equations for the wave vectors of bound energy levels of a finite barrier rectangular-shaped quantum dot, Eq. (36) in [21], are quite similar to ours—(22), (23), but somewhat more complicated. The ground state energy of electrons and holes in a core/shell quantum dot is given by Eq. (21) of [22], an equation similar to ours, just mentioned previously. Such results are important, inter alia, for the interpretation of photoluminescence spectra of heterojunctions.

Acknowledgements

The author acknowledges the financial support of the IFIN-HH-ANCSI project PN 16 42 01 01/ 2016 and of the IFIN-HH-JINR grant 04-4-1121-2015/2017.

Author details

Victor Barsan

Address all correspondence to: vbarsan@theory.nipne.ro

National Institute for Physics and Nuclear Engineering, Bucharest-Magurele, Romania

References

[1] von Roos O. Position-dependent effective masses in semiconductor theory. Physical Review. 1983;**B27**:7547-7559. DOI: 10.1103/PhysRevB.27.7547

[2] Lévai G, Özer O. An exactly solvable Schrödinger equation with finite positive position-dependenteffective mass. Journal of Mathematical Physics. 2010;**51**:092103(13 pp). DOI: 10.1063/1.3483716

[3] Nikitin AG, Zasadko TM. Superintegrable systems with position dependent mass. Journal of Mathematical Physics. 2015;**56**:042101(13 pp). DOI: 10.1063/1.4908107

[4] Sebawe Abdalla M, Eleuch H. Exact solutions of the position-dependent-effective mass Schrödinger equation. AIP Advances. 2016;**6**:055011(7 pp). DOI: 10.1063/1.4949567

[5] Bastard G. Wave mechanics applied to semiconductor heterostructures. Les Editions de Physique. 1990. 366 p

[6] Ihn T. Semiconductor Nanostructures. Oxford: Oxford University Press; 2011. 552 p

[7] Barsan V. New applications of the Lambert and generalized Lambert functions in ferromagnetism and quantum mechanics. arXiv:1611.01014v2. 2016

[8] de Alcantara Bonfim OF, Griffiths DJ. Exact and approximate energy spectrum for the finite square well and related potentials. American Journal of Physics. 2006;**74**:43-48. DOI: https://doi.org/10.1119/1.2140771

[9] Barsan V. Algebraic approximations for transcendental equations with applications in nanophysics. Philosophical Magazine. 2015;**95**:3023-3038. DOI: 10.1080/14786435.2015.1081425

[10] Song Y, Chen S. Janus nanoparticles: Preparation, characterization, and applications. Chemistry, an Asian Journal. 2014;**9**:418-424. DOI: 10.1002/asia.201301398

[11] Singh VA, Kumar L. Revisiting elementary quantum mechanics with the Daniel—Duke boundary conditions. American Journal of Physics. 2006;**74**:412-418. DOI: 10.1119/1.2174031

[12] Singh S, Pathak P, Singh VA. Approximate approaches to the one-dimensional finite potential well. European Journal of Physics. 2011;**32**:1701-1710. DOI: 10.1088/0143-0807/32/6/023

[13] Fluegge S. Practical Quantum Mechanics (I). Berlin: Springer-Verlag; 1971. 333 p. DOI: 10.1007/978-3-642-61995-3

[14] Pitkanen PH. Rectangular potential well problem in quantum mechanics. American Journal of Physics. 1955;**23**:111-113. DOI: 10.1119/1.1933912(1955)

[15] Barker BI, Rayborn GH, Ioup JJ, Ioup GE. Approximating the finite square well with an infinite well: Energies and eigenfunctions. American Journal of Physics. 1991;**59**:10381042

[16] Barsan V, Dragomir R. A new approximation for the square well problem. Optoelectronics and Advanced Materials. 2012;**6**:917-925

[17] Barsan V. A new analytic approximation for the energy eigenvalues of a finite square well. Romanian Reports in Physics. 2012;**64**:685-694

[18] Barsan V, Ciornei M-C. Semiconductor quantum wells with BenDaniel-Duke boundary conditions: Approximate analytical results. European Journal of Physics. 2017;**38**:015407 (22 pp). DOI: 10.1088/0143-0807/38/1/015407

[19] Weisstein EW. "Cubic Formula." From MathWorld—A Wolfram Web Resource. http://mathworld.wolfram.com/CubicFormula.html

[20] Biswas D, Kumar S, Das T. Interdiffusion induced changes in the photoluminescence of $In_xGa_{1-x}As/GaAs$ quantum dots interpreted. Journal of Applied Physics. 2007;**101**:026108 (3 pp). DOI: 10.1063/1.2430510

[21] Ata E, Demirhan D, Buyukkilic F. 2-d finite barrier rectangular quantum dots: Schroedinger description. Physica. 2014;**E62**:71-75. DOI: 10.1016/j.physe.2014.11.011

[22] Ibral A, Zouitine A, Assaid EM, Achouby HE. Polarization effects on spectra of spherical core/shell nanostructures: Perturbation theory against finite difference approach. Physica. 2015;**B458**:73-84. DOI: 10.1016/j.physb.2014.11.009

Section 2

Semiconductors for Energy Applications

Chapter 4

Nanostructured ZnO, Cu_2ZnSnS_4, $Cd_{1-x}Zn_xTe$ Thin Films Obtained by Spray Pyrolysis Method

Oleksandr Dobrozhan, Denys Kurbatov,
Petro Danilchenko and Anatoliy Opanasyuk

Additional information is available at the end of the chapter

http://dx.doi.org/10.5772/intechopen.72988

Abstract

The paper presents the investigation on the influence of substrate temperature T_s and the sprayed initial solution volume V_s on structural, substructural, optical properties, and elemental composition of ZnO and Cu_2ZnSnS_4 (CZTS) films as well as state-of-the-art of studying the $Cd_{1-x}Zn_xTe$ (CZT) films obtained by spray pyrolysis technique. The single-phase nanocrystalline ZnO films with average crystallite size of D_C = 25–270 nm and thickness of d = 0.8–1.2 μm can be deposited at substrate temperatures of T_s > 473 K. The continuous CZTS films with optimal thickness (d = 1.3 μm) for application as absorber layers in solar cells were deposited at the sprayed initial precursor volume of V_s = 5 ml. The increase of the substrate temperature up to 673 K caused the significant improvements in the stoichiometry of ZnO films. The optimal stoichiometry ratio of CZTS films for application in solar cells was obtained at V_s = 3–4 ml. Optical study of ZnO films showed that these films have a high-transmission coefficient values of T = 60–80%. To the best of our knowledge, there is the lack of works devoted to the study of CZT films obtained by spray pyrolysis technique.

Keywords: ZnO, Cu_2ZnSnS_4, $Cd_{1-x}Zn_xTe$, thin films, pulsed spray pyrolysis

1. Introduction

ZnO is an *n*-type direct-gap semiconductor with a wide band gap (E_g = 3.37 eV at T = 300 K) having the highest value of exciton energy (60 meV) among the binary compounds [1]. This material is a perspective for application in microelectronics, nanoelectronics, optoelectronics, sensors, solar cells among others due to its unique physical, electrical, and optical properties, non-toxic nature, and chemical and thermal stability in the ambient atmosphere [2]. It should be noted that, at present, Ukrainian sector of renewable energy, in particular solar, is developing rapidly. First of all it is made possible, thanks to the government support policy. In turn, it leads to an increased interest in the development of new solar cell designs to create further production of solar modules higher efficiency [3, 4]. Due to the absence of rare and toxic elements in zinc oxide compound and possibility to apply low-cost deposition techniques, this material may be an alternative to the traditional ITO ($(In_2O_3)_{0.9}$-$(SnO_2)_{0.1}$) and FTO (SnO_2:F)) transparent conductive layers in thin-film solar cells (SCs) and another optoelectronic device [5]. Nowadays, the perspective substitution of the traditional Si, CdTe, Cu(In,Ga)$(S,Se)_2$ absorption layers in thin-film SCs is considered Cu_2ZnSnS_4 (CZTS) semiconductor compound which has the optimal optical properties (E_g = 1.5 eV, $\alpha \sim 10^4$–10^5 cm^{-1}) [6].

$Cd_{1-x}Zn_xTe$ (CZT) solid solutions are perspective alternative absorption materials to Cu(In,Ga)$(S,Se)_2$ in the tandem solar cells having the band gap value of E_g = 1.1 eV. The appealing advantage of CZT compound is the variation of band gap by changing the zinc concentration. The optimal CZT solid solution with E_g ~1.7 eV can be obtained at the chemical composition of x ~ 0.2 [7]. To achieve the best working characteristics of devices, ZnO, CZTS, and CZT films must have the single-phase structure with large coherent domain sizes (CDS) L, low levels of microdeformations ε, microstresses σ, dislocation concentrations ρ, and well-controlled elemental composition. Unfortunately, typically these films have a high level of defects and secondary phases with different band gaps worsening the performance of the devices based on them. ZnO and CZT films in solar cells should possess high-transmission coefficients and controlled band gaps. Moreover, in order to improve the structural and optical properties of films for application in the low-cost optoelectronic devices, ZnO, CZTS, and CZT films should be deposited by low-cost, non-vacuum methods with optimized physical and technological deposition conditions.

Among the methods to deposit the ZnO, CZTS, and CZT films, special attention is paid to the spray pyrolysis technique having unique advantages: simplicity, efficiency, and cheapness. This technique provides the non-vacuum deposition of a large-area thin film with well-controlled properties.

It was shown [8, 9] that the greatest influence on physical properties and elemental composition of ZnO film has a substrate temperature T_s, CZTS film—the sprayed initial precursor volume V_s. It should be noted that until now CZT films deposited by spray pyrolysis technique are not well-studied, except some works [10, 11].

Thus, the investigation of the influence of deposition conditions on structural, substructural, and optical properties of ZnO, CZTS, and CZT films deposited by spray pyrolysis technique is the perspective in terms of its application in highly efficient optoelectronic devices.

2. ZnO, CZTS, and CZT thin films deposition methods. Peculiarities of the spray pyrolysis technique

The wide range of methods is well developed to deposit ZnO, CZT, and CZTS films which split into physical (for example, magnetron sputtering [12–14]) and chemical (for example, spray pyrolysis [5, 8–10]) techniques. Typically, the physical methods allow to obtain more perfect films with a higher structural quality, and these methods provide a precise control of the films thickness and low content of defects in deposited material compare to chemical methods, but physical deposition techniques require the usage of more complicated equipment and presence of high level of vacuum, thus they are energy-consuming. In contrary, chemical techniques to deposit ZnO, CZTS, and CZT films are low-cost and energy savers. Among them, spray pyrolysis method is considered as the most promising technique. This technique is simple and non-vacuum providing the deposition of the continuous, porous, nanostructured films, and multilayered structures [15].

Taking into account the increased interest to the nanosized materials with properties, significantly different to bulk materials (caused by quantum-size effect), several scientific groups have obtained the nanocrystalline ZnO and CZTS films [16, 17]. It is important to note that the works dedicated to the study of the nanosized structures used chemical techniques for films deposition. ZnO, CZTS, and CZT films deposited by spray pyrolysis technique are not yet well-studied; this fact conditioned the aim of our study.

The image of laboratory setup developed for the deposition of ZnO, CZTS, and CZT films by pulsed spray pyrolysis is showed in **Figure 1**. It consists of a spraying gun with initial precursor volume reservoir (1), spraying nozzle (2), and microcontroller block (3), allowing the control of the number of spraying cycles, time, and pauses between cycles. To the spraying gun, the compressor with pressure regulator (4) is connected with the aim of producing the air flow for transportation of the dispersed precursor onto heated substrate surface. Between the spraying gun and the compressor, an electromagnetic valve (5) is installed, where the "open" and "closed" regimes are controlled by the microcontroller block (3). The heating of substrate (6) is provided by the heating plate (7). During the deposition of films by spray pyrolysis technique, the properties of ZnO, CZTS, and CZT condensates are dependent on the precursor choice and physical, chemical deposition conditions. **Table 1** presents the overview of deposition conditions and precursors typically used to deposit the ZnO, CZTS, and CZT films by spray pyrolysis technique.

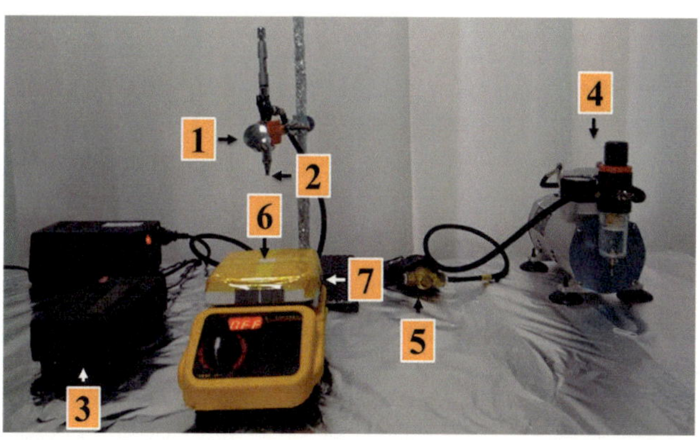

Figure 1. Image of the experimental setup for ZnO, CZTS, and CZT films deposition by pulsed spray pyrolysis: (1) spraying gun with initial precursor volume reservoir, (2) spraying nozzle, (3) microcontroller block, (4) compressor, (5) electromagnetic valve, (6) substrate, and (7) heating plate [18, 19].

№	Initial precursor	Solvent	Concentration (M)	Substrate type	Substrate temperature, T_s (K)	Ref.
ZnO films deposition						
1	Zinc chloride ($ZnCl_2$)	H_2O	0.10	Silicon	623–823	[21]
		H_2O	0.10	Glass	773	[22]
		H_2O	0.10	Glass	523–723	[23]
2	Zinc acetate ($Zn(CH_3COO)_2{\cdot}2H_2O$)	H_2O	0.04	Glass	573	[24]
3	Zinc acetate $Zn(CH_3COO)_2$	$H_2O + CH_3OH$	0.20	Glass	693	[25]
		H_2O	0.50	Glass	453–723	[26]
		H_2O	0.10	Glass	623	[27]
CZT films deposition						
4	Cadmium chloride ($CdCl_2$) Zinc chloride ($ZnCl_2$) Tellurium chloride ($TeCl_4$)	H_2O	0.02 (1:1:3)	Glass	250–325	[10, 11]
CZTS films deposition						
5	Copper chloride ($CuCl_2$) Zinc chloride ($ZnCl_2$) Tin chloride ($SnCl_2$) Thiourea ($CS(NH_2)_2$)	$(CH_3)_2SO$	0.010 0.005 0.005 0.040	Soda-lime glass	623	[28]
6	Copper chloride ($CuCl_2$) Zinc chloride ($ZnCl_2$) Tin chloride ($SnCl_2$) Thiourea ($CS(NH_2)_2$)	$H_2O + C_2H_5OH$	0.020 0.010 0.010 0.080	Soda-lime glass	553–633	[29]

Table 1. Precursors and physical-chemical conditions to deposit the ZnO, CZTS, and CZT films by spray pyrolysis method.

It should be noted that in order to obtain initial molecular solution for the deposition, the typical materials are metal salts dissolved in polar solvents, particularly in water, ethanol, etc. The most common substrates used are the non-oriented glass and silicon slides. The average substrate temperature is in the range of 250–823 K. It should be noted that these values are lower in comparison to the substrate temperatures used in physical methods.

3. Morphological, structural, and substructural properties of ZnO and CZTS films obtained by spray pyrolysis technique

The surface morphology, structural, substructural, optical properties, and chemical composition of ZnO and CZTS films deposited by spray pyrolysis method are determined by its physical, chemical, and technological deposition conditions.

3.1. The morphological properties

SEM images of ZnO films deposited at different substrate temperatures are presented in **Figure 2a–d** [19, 20]. It has been shown that at substrate temperatures higher than 473 K, crack-free and continuous nanocrystalline ZnO films with a good adhesion to substrate were formed.

The average grain size in the condensates was in the range of D_C = (25–270) nm (see inset in **Figure 2d**), increasing under the increase the deposition temperature up to 673 K. Whereas, the film thickness determined by the cross-sectional image was d = 0.8–1.2 μm.

One of the main film parameters of CZTS films is its thickness, which is typically controlled by the dispersed precursor volume V_s. Dependence CZTS films properties vs. films thickness in the range of d = 0.244–0.754 μm was studied by authors [9]. These values are not optimal for absorption of nearly 100% solar radiation because of the necessity CZTS films with thickness of d = 1–3 μm [6].

Thus, we have studied the CZTS films deposited by spray pyrolysis technique at different sprayed initial precursor volumes which had the higher thickness than studied in Ref. [9].

In **Figure 2e–h**, the SEM images of CZTS films and its cross-section deposited at different V_s are presented. It can be seen that in the range of studied values continuous films were formed,

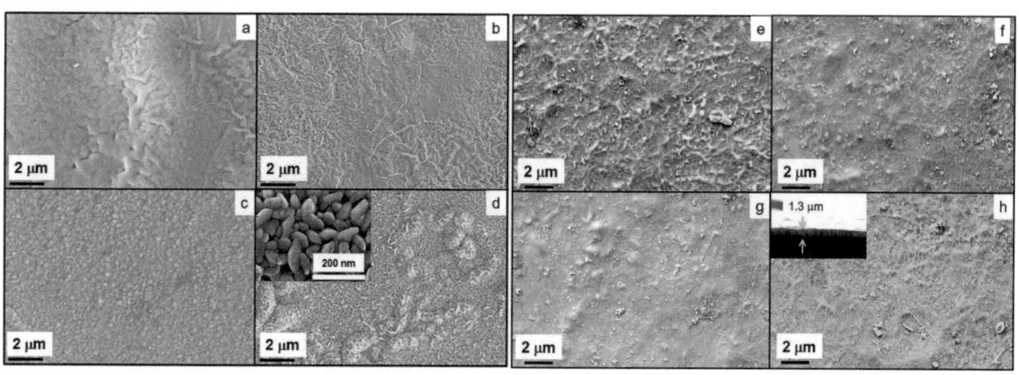

Figure 2. SEM images of ZnO films surface deposited at different temperature T_s, K: 473 (a), 573 (b), 623 (c), and 673 (d). Inset (d) shows the film surface with high resolution obtained at 673 K [19] and CZTS films surface deposited at different dispersed initial precursor volumes V_s, ml: 2 (e), 3 (f), 4 (g), and 5 (h). Inset (h) shows the films' cross-section [20].

which had a well adhesion to substrate and characterized by the absence of cracks and holes on their surfaces. The maximal layer thickness is determined by the cross-section method and it was d = 1.3 μm at V_s = 5 ml (**Figure 2h**).

3.2. Structural and substructural properties

Structural and substructural properties of ZnO and CZTS films have a significant influence on functional characteristics of devices [18–20]. Thus, its study is an important scientific objective. For example, band gap of zinc oxide films can be significantly increased by means of nanostructuring due to the quantum-size effects. At the same time, CZTS films as the absorber layers is SC should have the crystallites with sizes larger that diffusion length of minority charge carriers [6]. However, the films obtained by spray pyrolysis are usually characterized by high levels of microdeformations, microstresses, and density of dislocations in comparison to the values observed in the condensates deposited by physical vacuum methods, e.g., thermal evaporation, magnetron sputtering, etc.

The detailed description of methods applied to study structural, substructural, and optical properties of films is described elsewhere [18–20].

In **Figure 3a**, the XRD patterns of ZnO films deposited at different substrate temperatures are presented. On the diffraction patterns of the low-temperature films is dominated the diffraction line at angles 35.60–36.10° that corresponds to the reflection from (101) plane of ZnO hexagonal phase. On the diffraction patterns of the films deposited at T_s > 573 K, the lines at angles 31.80° and 34.80° are dominated, which correspond to the reflections of (100) and (002) crystallographic planes, respectively. X-ray analysis has been shown that deposited samples are single-phase and contain ZnO hexagonal phase. Secondary phases have not been determined by XRD analysis.

In **Figure 3b**, the X-ray patterns of CZTS films deposited at different dispersed solution volumes are shown. As can be seen from **Figure 3b**, on X-ray patterns is dominated the line on angles 28.05–28.50° which corresponds to the reflection from (112) CZTS tetragonal phase

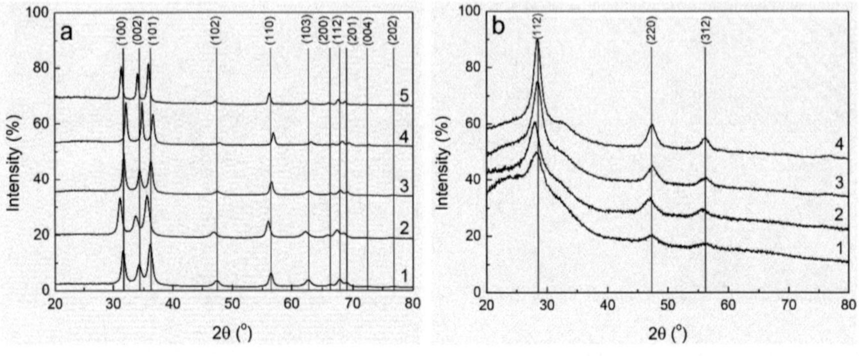

Figure 3. XRD patterns of ZnO films obtained at different substrate temperatures T_s, K: 473 (1), 523 (2), 573 (3), 623 (4), and 673 (5) (a); and CZTS films deposited at different dispersion solution volume V_s, ml: 2 (1), 3 (2), 4 (3), and 5 (4) (b). Vertical lines correspond to the JCPDS cards (ZnO— № 01–089-1397; CZTS— № 00–026-0575) [19, 20].

crystallographic plane. There are also presented lines at angles 47.15–47.50° and 55.55–56.45° which correspond to the reflection from (220) and (312) CZTS planes, respectively. It should be noted that during the increasing of precursor volume the intensity of peaks is increased and its half-width is decreased. It may be caused by the increasing of film thickness and by improvement of the films' crystalline quality. It is well-known that intensities ratio between the number of diffraction reflections from kesterite and stannite crystallographic planes is different [30]. Taking into account this fact, determination of these ratios gives an opportunity to estimate precisely the materials dominate phase. The measured intensity ratio $I_{(112)}/I_{(220)}$ from (112) and (220) crystallographic planes films was 2.23–2.56. These values are similar to the values, determined for un-doped films with kesterite phase ($I_{(112)}/I_{(220)} \approx 2.80$) [31]; thus, most probably, the investigated films have a kesterite phase. This conclusion is confirmed by the experimental measurements of lattice constants ratio ($c/2a$ = 0.9970–1.0203), that was similar to 1.0. These values are typical for kesterite [32].

Lattice parameter of the materials is a characteristic which is very sensitive to stoichiometry varying, impurities introduction, oxidation, etc. Thus, the precise determination of these values allows us to study the corresponding processes.

In **Figure 4**, the dependencies of ZnO and CZTS films lattice parameters *a*, *c* vs. deposition conditions are presented. In **Figure 4a**, it can be seen that during the increase of substrate temperature, measured parameters *a*, *c* for ZnO films are approached to the reference data that may be caused by the film stoichiometry improvement. High-temperature condensates composition approaching to the stoichiometric is confirmed by the chemical composition analysis data. In case of CZTS films (**Figure 4b**) most similar to the reference data *a* and *c* values are obtained by us at V_s = 4–5 ml, that is well-correlated with elemental composition analysis. It was estimated that lattice parameters are varied in the range of a_{ZnO} = 0.32477–0.32554 nm, c_{ZnO} = 0.51507–0.52111 nm, c/a_{ZnO} = 1.5822–1.6046, a_{CZTS} = 0.5423–0.5480 nm, c_{CZTS} = 1.0823–1.1182 nm, and $c/2a_{CZTS}$ = 0.9970–1.0203, the unit cell volume was in the range of $V_{cell(ZnO)}$ = 0.0427–0.0477 nm^3 and $V_{cell(CZTS)}$ = 0.3183–0.3358 nm^3, that is well-correlated to the reference data [31] and values obtained for ZnO and CZTS films deposited by spray pyrolysis technique in Ref. [34].

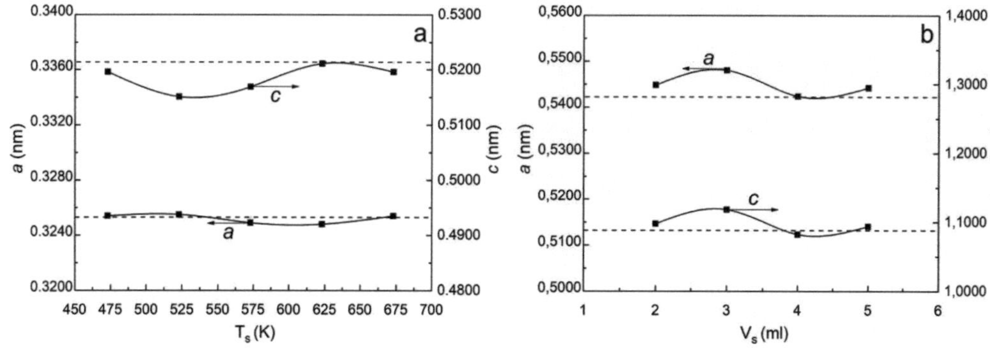

Figure 4. The dependencies of the lattice parameters *a*, *c* on the physical and technological deposition conditions [substrate temperature T_s, K—for ZnO films (a) dispersion solution volume V_s, ml—for CZTS films (b)]. Horizontal lines are corresponding to the stoichiometric material.

In **Figure 5**, the results of measurements L and ε parameters in studied films by threefold convolution method are presented. As can be seen from **Figure 5a**, in ZnO films during the increasing of substrate temperature from 473 to 673 K, there is a tendency of the CDS values increasing in direction of [100]—from $L \sim 14$ to ~ 21 nm, in direction of [101]—from $L \sim 11$ to ~ 20 nm and in direction of [102]—from $L \sim 10$ to ~ 63 nm. Similar L-T_s dependencies are observed in our previous works [33], where II-VI type compounds (CdTe, ZnS, ZnSe, and ZnTe) were obtained by close-spaced vacuum sublimation technique. At the same time, the microdeformations level in ZnO films in direction of [100] is decreased from $\varepsilon \sim 1.6 \times 10^{-5}$ to $\sim 0.5 \times 10^{-5}$, in direction of [101]—from $\varepsilon \sim 3.5 \times 10^{-5}$ to $\sim 1.2 \times 10^{-5}$, in direction of [102]—from $\varepsilon \sim 1.0 \times 10^{-5}$ to $\sim 0.7 \times 10^{-5}$, at the same substrate temperatures range (**Figure 5b**). Similar ε decreasing at substrate temperatures of $T_s > 573$ K is observed in CdTe and ZnTe films [35].

As can be seen from **Figure 5c**, in CZTS films during the increasing of the dispersion solution volume from 2 to 5 ml, CDS values are almost were not changed: $L \sim 24$–26 nm (for (112)–(220) planes pair), $L \sim 25$–27 nm (for(112)–(312) planes pair), and $L \sim 39$–40 nm (for (220)–(312) planes pair). Consequently, V_s varying influence on CDS sizes is negligible. It should be noted that obtained results of L measurements are well-correlated to the results presented in Refs. [36], where CZTS films were deposited under the similar experimental conditions. At the

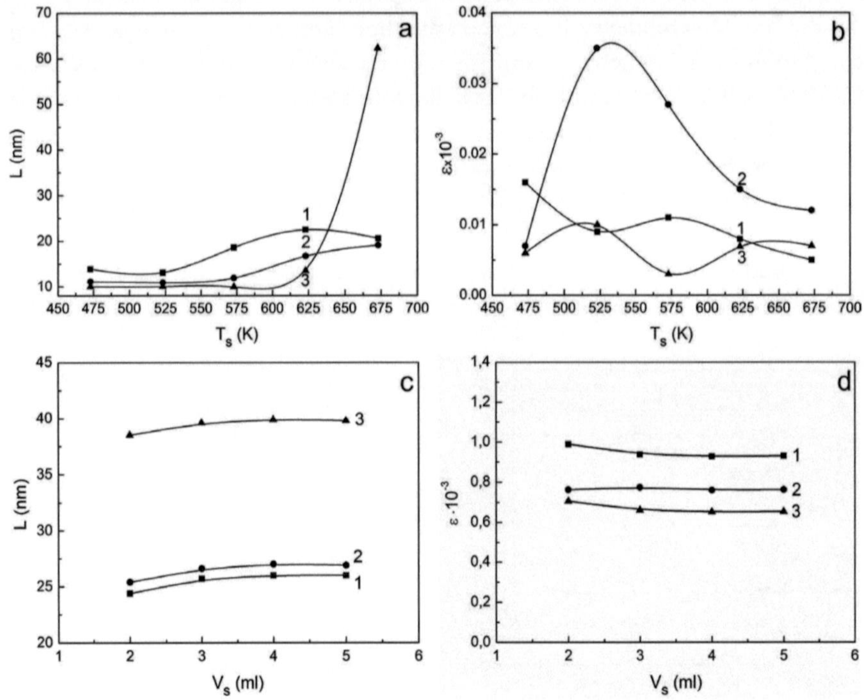

Figure 5. Influence of the substrate temperature on CDS values (a) and microdeformations level (b) of ZnO films on direction normal to the (100)–(200) (1), (101)–(202) (2), (102)–(103) (3) crystallographic planes and of the dispersion solution volume V_s on L (c) and ε (d) for CZTS films on direction normal to the (112)–(220) (1), (112)–(312) (2), (220)–(312) (3) planes. The threefold convolution technique was used.

same time, the microdeformations level in CZTS films for the directions normal to (112)–(220) crystallographic planes is varied in the range of $\varepsilon \sim (0.93–0.99) \times 10^{-3}$; for (112)–(312) planes—$\varepsilon \sim (0.76–0.77) \times 10^{-3}$; for (220)–(312) planes—$\varepsilon \sim (0.65–0.71) \times 10^{-3}$ (**Figure 5d**). It should be noted that measured microdeformations values in CZTS films are lower than presented in Ref. [37], where $\varepsilon \sim (1.26–6.60) \times 10^{-3}$. The measured level of microdeformations allowed us to determine the level of microstresses (σ) in nanocrystalline ZnO and CZTS films. It was estimated that microstresses levels in ZnO and CZTS films varied in the ranges of σ_{ZnO} = 0.48–1.53 MPa, σ_{CZTS} = 5.2–20.3 MPa, respectively. The influence of T_s onto σ level in ZnO films was also studied in Refs. [8], where authors estimated that during the increase of T_s from 623 to 723 K the compression stress level σ was decreased (1.77–1.47 GPa).

In CZTS films, during the increase of the dispersion solution volume microstresses level is decreased, wherein the smallest σ values are observed in layers obtained at V_s = 5 ml.

In **Figure 6**, the results of measurements of the dislocations concentration on the boundaries (ρ_L) and within volume (ρ_ε) CSDs blocks and of the general dislocations concentration (ρ) in direction normal to (100) crystallographic plane for ZnO films or (112) crystallographic plane for CZTS films are presented. Studied ZnO layers are characterized by rather low values of $\rho = (1.3–6.1) \times 10^{13}$ lines/m^2 compare to the results obtained by other authors. As it can be seen from **Figure 6a**, during the increase of T_s, there is a tendency to decrease ρ values. In Ref. [38], authors have estimated that in ZnO nanocrystalline films with thickness of d = 0.135–0.392 μm, deposited at T_s = 473 K, the dislocation concentration values are higher ($\rho = (1.29–4.15) \times 10^{15}$ lines/m^2) than determined here. In Ref. [39], authors have also obtained higher values ($\rho = (2.4–5.8) \times 10^{13}$ lines/m^2) compared to our results. It has been shown that during the increase of V_s (**Figure 6b**) in CZTS films general dislocation density ρ does not almost change in all investigated directions.

The smallest values of $\rho = (17.0–19.3) \times 10^{15}$ lines/m^2 are obtained in case of film deposited at dispersion volume of V_s = 5 ml. It should be noted that these values ρ are smaller than observed earlier for CZTS films deposited by chemical methods (spray pyrolysis – $\rho = (11.6–80.3) \times 10^{16}$ lines/m^2) [35]), and higher compare to the films obtained by vacuum methods (thermal evaporation – $\rho = (0.64–4.00) \times 10^{14}$ lines/m^2 [40]).

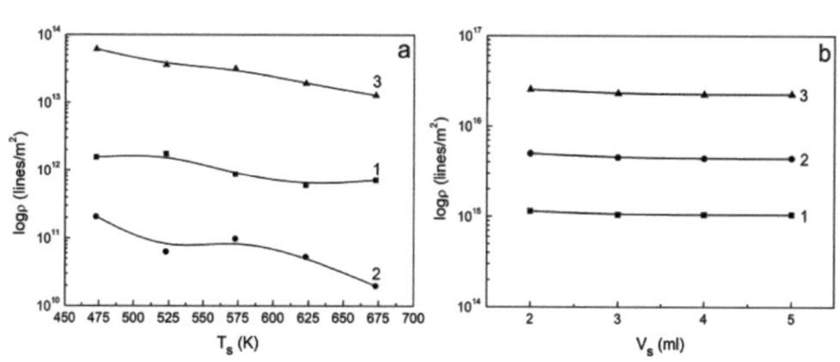

Figure 6. The influence of substrate temperature T_s (in case of ZnO films (a)) and dispersed initial solution volume V_s (in case of CZTS films (b)) on dislocations density ρ: on the sub-grain boundaries (1), within CDS units (2) and general dislocations concentration (3) for the direction normal to (100)–(200) planes for ZnO and to (112)–(220) planes for CZTS. The measurements error was varied in the range of 15–20%.

3.3. The study of the stoichiometry

Energy dispersed analysis of the X-ray spectra (EDAX) allows us to determine the elemental composition of ZnO and CZTS films obtained in present work. Results determined for films deposited at different physical-chemical and technological conditions are presented in **Table 2**. As it can be seen, ZnO films have some oxygen surplus on zinc. Besides, films stoichiometry is increased during the increasing of the substrate temperature. This fact is confirmed by the concentration ratios C_O/C_{Zn}, that are parts of compound ($\gamma_{ZnO} = 1.4 - T_s = 473$ K, $\gamma_{ZnO} = 1.2 - T_s = 623$ K). The impurities connected to the films contamination by the precursor materials have not been determined.

The control of CZTS films elemental composition is a complex and important task because of its probable determination of the phase conditions, crystal structure, optical, and electrical properties of investigated layers. It was estimated that in CZTS films some copper, zinc, and tin are present in surplus and has some sulfur deficiency. Sulfur losses in films during the pyrolytic reaction of the initial precursor near the surface of the heated substrate may be caused by its high volatility [41]. It should be noted that stoichiometry of studied films is some improved during the increasing of dispersed precursor volume. Also, the obtained ratio γ_{CZTS_1} = (0.80–0.84) in CZTS films deposited at precursor dispersion with volume of V_s = 2–3 ml is close to the optimal values necessary to develop SCs with high solar energy conversion efficiency (γ_{CZTS_1} = (0.8–0.9), γ_{CZTS_2} = (1.1–1.2)) [40, 42]. For film obtained by dispersion precursor volume of 3 ml for this requirement corresponds the next ratio—γ_{CZTS_2} = 1.2. Impurities related to the films' contamination by the precursor's materials have also not been observed in CZTS layers.

ZnO							
T_s (K)	C_{Zn} (at. %)	C_O (at. %)			γ_{ZnO}		
473	41.8	58.2			1.4		
523	42.3	57.7			1.4		
573	42.6	57.4			1.3		
623	44.3	55.7			1.2		
673	44.0	56.0			1.2		
Stoichiometry	50.0	50.0			1.0		
CZTS							
V_s (ml)	C_{Cu} (at. %)	C_{Zn} (at. %)	C_{Sn} (at. %)	C_S (at. %)	γ_{CZTS_1}	γ_{CZTS_2}	γ_{CZTS_3}
2	28.6	21.4	14.3	35.8	0.8	1.5	0.8
3	27.0	17.3	14.7	40.8	0.8	1.2	0.7
4	27.7	16.3	15.1	40.9	0.9	1.0	0.6
5	26.4	15.2	15.4	43.0	0.9	1.0	0.6
Stoichiometry	25.0	12.5	12.5	50.0	1.0	1.0	0.5

Table 2. Measurement results of the chemical composition for ZnO and CZTS films obtained at different conditions.

http://dx.doi.org/10.5772/intechopen.72988

4. Optical properties of ZnO and CZTS films obtained by spray pyrolysis technique

4.1. Optical properties

The study and control of the optical properties of ZnO, CZTS, and CZT films is an important task with the aim of their usage in optoelectronic devices, especially for SCs development. It is well-known that optical characteristics of these films heavily dependent on morphological, structural, substructural properties, chemical composition, and physical (chemical) and technological deposition conditions.

In present work, the transmission light coefficient of ZnO films was in the range of $T = 60–80\%$ at the wavelength range of $\lambda = 430–800$ nm. The highest transmission values had films obtained at $T_s = 673$ K. It was estimated that measurement E_g values for ZnO films were in the range of 3.18–3.30 eV and were also dependent on T_s.

As can be seen from **Figure 7a**, band gap E_g of zinc oxide during the increasing of the deposition temperature is at first increased and in further decreased. This complex dependence of E_g may be caused by increasing of the grain sizes in films and by improvement of their structural quality during the increasing of T_s. It is well known [43] that in nanocrystalline films ($D_C < 100$ nm) band gap is determined by quantum effects, that leads to the increasing of E_g compare to the values observed in bulk materials. During the increasing of the grain size, quantum effects are gradually decreased. At the same time, due to the high level of the substructural defects (primarily dislocations) in nanocrystalline films, that have been given the local deformations on the materials lattice, its average E_g have been smaller than in bulk materials [16]. At high substrate temperatures, films with sufficient large grain size and low structural defect concentration were formed. As a result, the band gap of semiconductor is approaching the bulk value. Similar tendencies of E_g changing depending on the deposition temperature were observed in Refs. [44].

In **Figure 7b**, dependence of the materials E_g on the dispersed solution precursor volume V_s is presented. It should be noted that the smallest α values have been obtained for layers

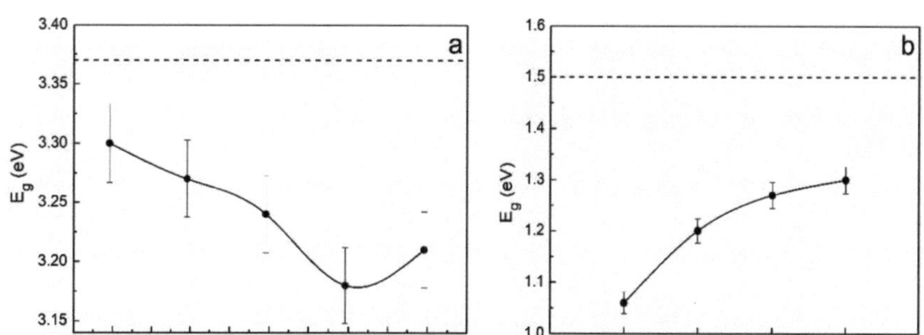

Figure 7. Bang gap (E_g) dependencies on substrate temperature T_s in ZnO films (a) and on dispersion solution volume V_s in CZTS films (b). Dashed line corresponds to band gap value in bulk ZnO ($E_g = 3.37$ eV) and bulk CZTS ($E_g = 1.50$ eV).

deposited at volume of $V_s = 2$ ml, the highest values—at $V_s = 5$ ml, respectively. It is quite typical because of the smallest and highest thickness values of the corresponding layers. During the increasing of dispersed initial precursor solution volume, the values of band gap were varied in the range of E_g = 1.06–1.30 eV and were approximately approached to the values typical for bulk stoichiometric material (E_g = 1.5 eV). It indicates on increasing of grain sizes and decreasing of films deficiency during the increasing of their thickness. Similar tendencies have been observed in Ref. [9].

4.2. Raman and Fourier transform IR (FTIR) spectra

Raman spectroscopy is an additional to X-ray diffraction analysis method of studying the phase composition and quality of ZnO, CZTS, and CZT thin films.

Raman spectra of ZnO films measured in the range of frequencies 90–800 cm^{-1} are presented in **Figure 8a**. In spectra, a number of different intensity lines on the next frequencies: 95–98, 333–336, 415, 439–442, 572, and 578–587 cm^{-1} are observed. Using the reference data, these lines were interpreted by us as the next phonon modes: $E_2^{low}(Zn)$ [43–45], E_2^{high}-E_2^{low} [46], $E_1(TO)$ [45], $E_2^{high}(O)$ [43–47], $A_1(LO)$ [43] and $E_1(LO)$ [45–46]. In **Figure 8a**, two intensive peaks, which correspond to E_2 mode, are also observed: peak E_2^{high}, which is relative to the oxygen anions, is localized at frequency of 439–442 cm^{-1} and peak E_2^{low}, which is relative to zinc cations, is localized at frequency of 95–98 cm^{-1}. It is well known [49] that the crystalline quality of ZnO films has a direct influence on the mode E_2 intensity. Besides, $E_2^{high}(O)$ peak is very sensitive to the presence of inner defects of material. The deviation of the frequency $E_2^{high}(O)$ peak from the value typical for bulk ZnO (437 cm^{-1}), that is observed by us in low-temperature condensates, indicates about the presence in zinc oxide high level of microstresses and stretched defects (dislocations) density of the lattice. It should be noted that during the increase of the substrate temperature, the $E_2^{high}(O)$ peak position is some red-shifted from 442 cm^{-1} to the typical bulk ZnO values—439 cm^{-1}, which indicates the decrease of σ and ρ levels.

FTIR spectroscopy is an addition to X-ray diffraction analysis and Raman spectroscopy technique, which allows to obtain an information about the elemental composition of the studied material and its contamination by the precursor impurities. The number of frequencies,

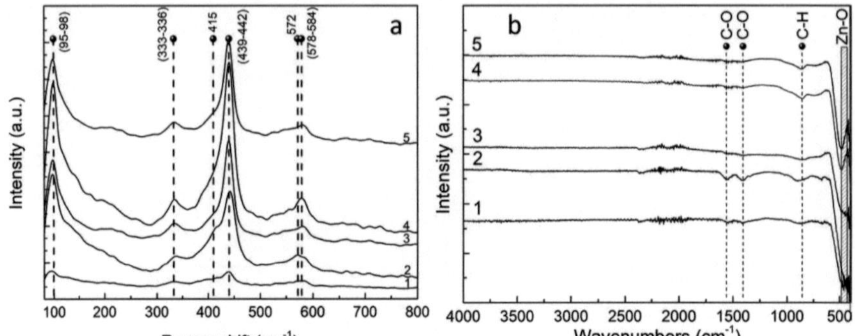

Figure 8. Raman (a) and FTIR (b) spectra of ZnO films deposited at different substrate temperatures T_s, K: 473 (1), 523 (2), 573 (3), 623 (4), and 673 (5).

where the light absorption and transmission in films are performed, allows us to determine the functional links between chemical elements which are part of the studied materials.

In **Figure 8b**, FTIR reflection spectra of ZnO films deposited at different substrate temperatures are presented. Although that thin films were deposited in air by chemical technique obtained spectra were comparatively pure.

At low frequencies (460–475 cm^{-1}), there is observed minima, which due to the reference data [48], correspond to Zn-O vibrational mode. It should be noted that FTIR spectra obtained on films deposited in all range of substrate temperatures have a C-Cl vibrational mode [50]. The presence of this connection may be caused by the usage of HCl acid, which was added as a precursor during its preparation. The acid paths are also observed in films. In FTIR spectra of ZnO films deposited at $T_s < 573$ K, peaks on the frequencies 1405 and 1560 cm^{-1} are presented; they were interpreted by us as symmetric and asymmetric C-O vibrational modes [50]. The absence of C-O connections in films deposited at $T_s > 573$ K indicates about the total precursor decomposition near the substrate surface at these temperatures. It eliminates the possibility of adsorption of the acetate elements on ZnO films surface during the pyrolysis, and it leads to the formation of single-phase zinc oxide polycrystalline films.

It is well known that in CZTS films, the presence of secondary phases, such as Cu_xS_y, Zn_xS_y, Sn_xS_y, Cu_xSnS_y, ZnO, and Zn_xSnO_y, is available [39, 51–53]. They are characterized by affiliated lattices, and they indicate on XRD patterns refractions on similar angles. It complicates the phase analysis by XRD technique. Thus, for precise identification of the secondary phases in CZTS compound, the researchers often use Raman spectroscopy in addition to XRD analysis [54]. It allows to identify not only secondary phases, but also kesterite and stannite. In **Table 3**,

Experimental data				**Literature data**			
V_s, ml				Raman shift, cm^{-1}	Symmetry	Mode	Reference
2	3	4	5				
Raman shift (cm^{-1})							
Green-laser (λ = 514.5 nm)							
142				143–144	E	*CZTS E*	[56]
340				338–339	A	*CZTS A*	[54]
664				672	A	*2a CZTS A*	[56]
Red laser (λ = 632.8 nm)							
339				338–339	A	*CZTS A*	[52]
663				672	A	*2a CZTS A*	[57]
UV-laser (λ = 325 nm)							
340				341	A	*CZTS A*	[56]
—	560	—	—	541	—	*ZnO*	[45]
664				672	A	*2a CZTS A*	[57]

Table 3. Peaks interpretation presented on Raman spectra of CZTS films.

the results of study the Raman spectra of CZTS films using as an excitation source the radiation of several lasers are presented. At all spectra regardless on the precursor volume and excitation laser type, the main peak on frequencies of (339–340) cm^{-1} is presented. It is well correlated to the results of previous studies [52, 54, 55]. In Raman spectra obtained using the green laser, lines on the next frequencies: 142, 340, and 664 cm^{-1} are also observed, which correspond to *CZTS E*, *CZTS A*, and *2a CZTS A* (*CZTS A* mode phonon replica) phonon modes, respectively [54–56].

Usage of the red- and UV-lasers as phonons excitation source allows us to increase the method's sensitivity onto the revealing of compounds with optical band gap close to E_g ~ 1.96 and ~ 3.81 eV (excitation radiations energies of corresponding lasers). On spectra, obtained using the red- and UV-lasers, are presented lines on frequencies 339–340 cm^{-1}, 663–664 cm^{-1} which correspond to the *CZTS A* and *2a CZTS A* phonon modes [52, 56, 57]. The usage of UV-laser in one of the studied films revealed a negligible number of ZnO secondary phase. In addition, these results are supported by the phonon excitation in Raman spectra on the frequency at 560 cm^{-1} on film obtained from the precursor volume dispersion of 3 ml. Other secondary phases in studied films are not revealed. Raman spectra of CZT films measured during the influence of green-laser excitation radiation (λ = 514 nm, E = 2.41 eV) are presented in **Figure 9**. In the spectra of the CZT sample (x = 0.32), peak which corresponds to LO_2(ZnTe) mode is observed. In these spectra, intensive peaks which correspond to A_1(Te) E_{TO}(Te) telluric modes are also detected. In the spectra of the CZT sample (x = 0.75), is observed a weak peak that corresponds to A_1(Te) mode, peaks of the next modes: LO_1(CdTe), TO_1(CdTe), TO_2(ZnTe), LO_2(ZnTe), and also detected the LO_2(ZnTe) mode resonant replica.

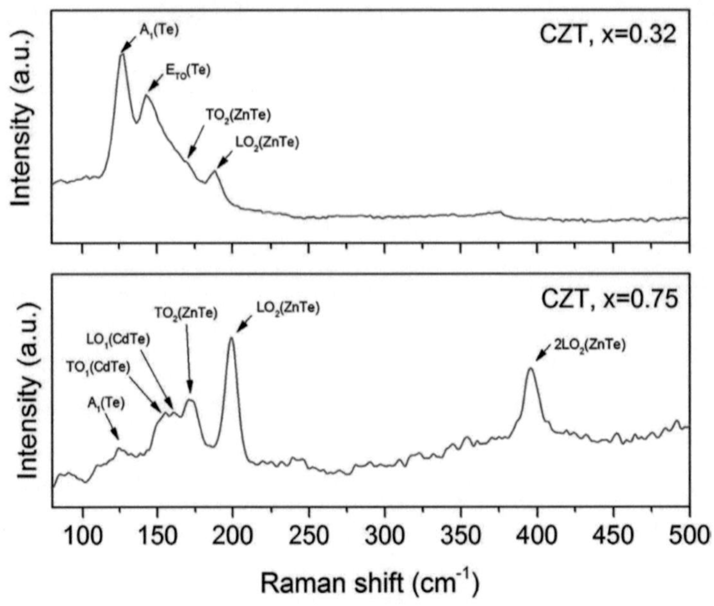

Figure 9. Raman spectra of CZT films measured during the impact of excitation irradiation of the wavelength 785 nm at room temperature (RT) [7].

5. Conclusions

As a results of the complex study of structure, substructure, optical properties, and elemental composition of ZnO, CZTS, and CZT films obtained by pulsed spray pyrolysis technique dependent on the physical (chemical) and technological deposition conditions, it was determined that ZnO nanocrystalline films have an average grain size of D_C = 25–270 nm and their thickness was d = 0.8–1.2 μm, and were formed at T_s > 473 K. CZTS continuous films with optimal thickness of d = 1.3 μm were deposited at dispersed initial precursor volume V_s = 5 ml. It was found that ZnO and CZTS films were polycrystalline in nature, single-phase, and had hexagonal and tetragonal phases, respectively. CZTS samples had a kesterite structure.

It has been shown that in ZnO during the increasing of substrate temperature there is a tendency to the increasing of the CDS; however, in CZTS films, their CSD values were weakly depended on the dispersed solution volume.

Lattice parameters values in ZnO and CZTS films deposited at T_s = 623 K, V_s = 4 ml were well-correlated to the reference data that confirms their optimal stoichiometry and crystalline quality.

It has been estimated that during the increase of T_s the microdeformations level, microstresses, and dislocation density in ZnO films were decreased; in CZTS films, these parameters were weakly dependent on V_s.

It has been determined that during the increasing of substrate temperature to 623 K stoichiometry of ZnO layers was improved (γ_{ZnO} = 1.2). It has been shown that optimal for usage in SCs CZTS films, their stoichiometry ratios γ_{CZTS_1} = 0.8–0.9, γ_{CZTS_2} = 1.1–1.2, γ_{CZTS_3} = 0.7 were obtained at V_s = 3–4 ml.

Study of the optical characteristics of ZnO films allow to estimate the high values of transmission coefficient T = 60–80%. Measured E_g values of ZnO layers were determined in the range of 3.18–3.30 eV and had a complex dependence on T_s. During the increase of V_s, the values of E_g = 1.06–1.30 eV of CZTS layers were approximately approached to the reference data E_g = 1.5 eV. Raman spectra analysis of ZnO films confirmed the results of the XRD study, namely decreasing ε, σ, and ρ values during the increase of T_s. CZTS films' Raman spectra analysis has confirmed the single-phase nature of condensates. FTIR study indicated the absence of precursor impurities in ZnO films obtained at T_s > 573 K.

CZT film spectra (x = 0.32) had a mode LO_2(ZnTe). In these spectra, intensive peaks corresponded to A_1(Te) and E_{TO}(Te) tellure modes were also determined. CZT film spectra (x = 0.75) have a weak mode A_1(Te), peaks of LO_1(CdTe), TO_1(CdTe), TO_2(ZnTe), and LO_2(ZnTe) modes, and also LO_2(ZnTe) mode resonant replica.

The results of a research study of the ZnO, CZTS, and CZT thin films will be used for the development of the devices, primarily, in third generation high-efficiency thin-film solar cells.

Acknowledgements

This work was supported by the Ministry of the Education and Science of Ukraine (Grants numbers: 0116U002619, 0115U000665c, 0116U006813, and 0117U003929).

Author details

Oleksandr Dobrozhan, Denys Kurbatov*, Petro Danilchenko and Anatoliy Opanasyuk

*Address all correspondence to: dkurbatov@sumdu.edu.ua

Department of Electronics and Computer Technology, Sumy State University, Sumy, Ukraine

References

[1] Ozgur U, Alilov Y, Teke A. A comprehensive review of ZnO materials and devices. Applied Physics Reviews. 2005;**98**:041301. DOI: 10.1063/1.1992666

[2] Look D. Recent advances in ZnO materials and devices. Materials Science and Engineering B. 2001;**80**:383-387. DOI: 10.1016/S0921-5107(00)00604-8

[3] Kurbatova T, Khlyap H. State and economic prospects of developing potential of non-renewable and renewable energy resources in Ukraine (Review). Renewable and Sustainable Energy Reviews. 2015;**52**:217-226. DOI: 10.1016/j.rser.2015.07.093

[4] Kurbatova T, Khlyap H. GHG emissions and economic measures for low carbon growth in Ukraine. Carbon Management. 2015;**6**:7-17. DOI: 10.1080/17583004.2015.1065376

[5] Mahajan C, Takwale M. Intermittent spray pyrolytic growth of nanocrystalline and highly oriented transparent conducting ZnO thin films: Effect of solution spray rate. Journal of Alloys and Compounds. 2014;**584**:128-135. DOI: 10.1016/j.jallcom.2013.08.136

[6] Ito K, editor. Copper Zin Tin Sulfide-Based Thin Film Solar Cells. Chichester: Wiley; 2015. 440p. ISBN: 978-1-118-43787-2

[7] Kosyak V, Znamenshchykov Y, Cerskus A. Composition dependence of structural and optical properties of $Cd_{1-x}Zn_xTe$ thick films obtained by the close-spaced sublimation. Journal of Alloys and Compounds. 2016;**682**:543-551. DOI: 10.1016/j.jallcom.2016.05.065

[8] Prasada Rao T, Santhosh M, Safarulla K. Physical properties of ZnO thin films deposited at various substrate temperatures using spray pyrolysis. Physica B: Condensed Matter. 2010;**405**:2226-2231. DOI: 10.1016/j.physb.2010.02.016

[9] Shinde N, Deokate R, Lokhande C. Properties of spray pyrolysis deposited Cu_2ZnSnS_4 (CZTS) thin films. Journal of Analytical and Applied Pyrolysis. 2013;**100**:12-16. DOI: 10.1016/j.jaap.2012.10.018

[10] Gaikwad S, Tembhurkar Y, Dudhe C. Effect of substrate temperature on optical band gap and thickness of spray pyrolitically deposited $CdZnTe_2$ thin films. International Journal of Science and Research. 2017;**6**:1627-1634. DOI: 10.21275/ART20177528

[11] Gaikwad S, Tembhurkar Y, Dudhe C. Optical and electrical properties of spray pyrolitically deposited $CdZnTe_2$ thin films. International Journal of Pure and Applied Physics. 2017;**13**:231-240

[12] Arakelova E, Khachatryan A, Kteyan A. ZnO film deposition by DC magnetron sputtering: Effect of target configuration on the film properties. Thin Solid Films. 2016;**612**:407-413. DOI: 10.1016/j.tsf.2016.06.030

[13] Zhao Q, Hao R, Liu S. Fabrication and characterization of Cu_2ZnSnS_4 thin films by sputtering a single target at different temperature. Physica B: Condensed Matter. 2017;**523**:62-66. DOI: 10.1016/j.physb.2017.08.035

[14] Gao X, Zhu X, Sun H. Preparation and characterization of CdZnTe multilayer films by repeated RF magnetron sputtering. Journal of Materials Science: Materials in Electronics. 2017;**28**:4467-4474. DOI: 10.1007/s10854-016-6079-8

[15] Mooney J, Radding S. Spray pyrolysis processing. Annual Review of Materials Research. 1982;**12**:81-101. DOI: 10.1146/annurev.ms.12.080182.000501

[16] Tan S, Chen B, Sun X. Blueshift of optical band gap in ZnO thin films grown by metal-organic chemical-vapor deposition. Journal of Applied Physics. 2005;**98**:013505. DOI: 10.1063/1.1940137

[17] Riha S, Fredrick S, Sambur J. Photoelectrochemical characterization of nanocrystalline thin-film Cu_2ZnSnS_4 photocathodes. ACS Applied Materials and Interfaces. 2011; **3**:013505. DOI: 10.1021/am1008584

[18] Dobrozhan A, Opanasyuk A, Kolesnyk M. Substructural investigations, Raman, and FTIR spectroscopies of nanocrystalline ZnO films deposited by pulsed spray pyrolysis. Physica Status Solidi. 2015;**212**:2915-2921. DOI: 10.1002/pssa.201532324

[19] Dobrozhan O, Kurbatov D, Opanasyuk A. Influence of substrate temperature on the structural and optical properties of crystalline ZnO films obtained by pulsed spray pyrolysis. Surface and Interface Analysis. 2015;**47**:601-606. DOI: 10.1002/sia.5752

[20] Dobrozhan O, Loboda V, Ya Z. Structural and optical properties of Cu_2ZnSnS_4 films obtained by pulsed spray pyrolysis. Journal of Nano- and Electronic Physics. 2017; **9**:01028. DOI: 10.21272/jnep.9(1).01028

[21] Gorbik P, Dubrovin I, Filonenko M. Chemical method of obtaining the nanocrystalline texture ZnO films. Physics and Chemistry of Solid State. 2004;**5**:552-556

[22] Bacaksiz E, Aksu S, Yilmaz S. Structural, optical and electrical properties of Al-doped ZnO microrods prepared by spray pyrolysis. Thin Solid Films. 2010;**518**:4076-4080. DOI: 10.1016/j.tsf.2009.10.141

[23] Zaier A, Oum El az F, Lakfif F. Effects of the substrate temperature and solution molarity on the structural opto-electric properties of ZnO thin films deposited by spray pyrolysis. Materials Science in Semiconductor Processing. 2009;**12**:207-211

[24] Ayouchi R, Leinen D, Martin F. Preparation and characterization of transparent ZnO thin films obtained by spray pyrolysis. Thin Solid Films. 2003;**426**:68-77

[25] Ashour A, Kaid M, El-Sayed N. Physical properties of ZnO thin films deposited by spray pyrolysis technique. Applied Surface Science. 2006;**252**:7844-7848. DOI: 10.1016/j.apsusc.2005.09.048

[26] Riad A, Mahmoud S, Ibrahim A. Structural and DC electrical investigations of ZnO thin films prepared by spray pyrolysis technique. Physica B: Condensed Matter. 2001;**296**:319-325. DOI: 10.1016/S0921-4526(00)00571-8

[27] Shinde S, Patil G, Kajele D. Synthesis of ZnO nanorods by spray pyrolysis for H_2S gas sensor. Journal of Alloys and Compounds. 2012;**528**:109-114. DOI: 10.1016/j.jallcom.2012.03.020

[28] Moreno R, Ramirez E, Guzman G. Study of optical and structural properties of CZTS thin films grown by co-evaporation and spray pyrolysis. Journal of Physics Conference Series. 2016;**687**:012041. DOI: 10.1088/1742-6596/687/1/012041

[29] Nakayama N, Ito K. Sprayed films of stannite Cu_2ZnSnS_4. Applied Surface Science. 1996;**92**:171-175. DOI: 10.1016/0169-4332(95)00225-1

[30] Babu G, Kumar Y, Bhaskar P. Growth and characterization of co-evaporated Cu_2ZnSnS_4 thin films for photovoltaic applications. Journal of Physics D: Applied Physics. 2008;**41**:205305. DOI: 10.1088/0022-3727/41/20/205305

[31] Cui Y, Zuo S, Jiang J. Synthesis and characterization of co-electroplated Cu_2ZnSnS_4 thin films as potential photovoltaic material. Solar Energy Materials and Solar Cells. 2011;**95**:2136-2140. DOI: 10.1016/j.solmat.2011.03.013

[32] Paier J, Asahi R, Nagoya A. Cu_2ZnSnS_4 as a potential photovoltaic material: A hybrid Hartree-Fock density functional theory study. Physical Review B. 2009;**79**:115126. DOI: 10.1103/PhysRevB.79.115126

[33] Selected Powder Diffraction Data for Education and Training (Search Manual and Data Cards). Pennsylvania: International Center for Diffraction Data; 1998

[34] Rajeshmon V, Kartha C, Vijayakumar K. Role of precursor solution in controlling the opto-electronic properties of spray pyrolysed Cu_2ZnSnS_4 thin films. Solar Energy. 2012;**85**:249-255. DOI: 10.1016/j.solener.2010.12.005

[35] Opanasyuk A, Kurbatov D, Kosyak V. Characteristics of structure formation in zinc and cadmium chalcogenide films deposited on nonorienting substrates. Crystallography Reports. 2012;**57**:927-933

[36] Valdes M, Santoro G, Vazques M. Spray deposition of Cu_2ZnSnS_4 thin films. Journal of Alloys and Compounds. 2014;**585**:776-782. DOI: 10.1016/j.jallcom.2013.10.009

[37] Khalate S, Kate R, Kim J. Effect of deposition temperature on the properties of Cu_2ZnSnS_4 (CZTS) thin films. Superlattices and Microstructures. 2017;**103**:335-342. DOI: 10.1016/j.spmi.2017.02.003

[38] Ozutok F, Demirselcuk B, Sarica E. Study of ultrasonically sprayed ZnO films: Thermal annealing effect. Acta Physica Polonica A. 2012;**121**:53-55

[39] Kumar Y, Babu G, Bhaskar P. Preparation and characterization of spray-deposited Cu_2ZnSnS_4 thin films. Solar Energy Materials and Solar Cells. 2009;**93**:1230-1237. DOI: 10.1016/j.solmat.2009.01.011

[40] Touatia R, Ben Rabeh M, Kanzari M. Structural and optical properties of the new absorber Cu_2ZnSnS_4 thin films grown by vacuum evaporation method. Energy Procedia. 2014;**44**:44-51. DOI: 10.1016/j.egypro.2013.12.008

[41] Tanaka K, Moritake N, Uchiki H. Preparation of Cu_2ZnSnS_4 thin films by sulfurizing sol-gel deposited precursors. Solar Energy Materials and Solar Cells. 2007;**13**:1199-1201. DOI: 10.1016/j.solmat.2007.04.012

[42] Mitzi D, Gunawan O, Todorov T. The path towards a high-performance solution-processed kesterite solar cell. Solar Energy Materials and Solar Cells. 2011;**95**:1421-1436. DOI: 10.1016/j.solmat.2010.11.028

[43] Wong E, Searson P. ZnO quantum particle thin films fabricated by electrophoretic deposition. Applied Physics Letters. 1999;**74**:2939-2941. DOI: 10.1063/1.123972

[44] Ayouchi R, Martin F, Leinen D. Growth of pure ZnO thin films prepared by chemical spray pyrolysis on silicon. Journal of Crystal Growth. 2003;**247**:497-504. DOI: 10.1016/S0022-0248(02)01917-6

[45] Bendall J, Visimberga G, Szachowicz M. An investigation into the growth conditions and defect states of laminar ZnO nanostructures. Journal of Materials Chemistry. 2008;**18**:5259-5266. DOI: 10.1039/B812867G

[46] Shinde S, Bhosale C, Rajpure K. Structural, optical, electrical and thermal properties of zinc oxide thin films by chemical spray pyrolysis. Journal of Molecular Structure. 2012;**1021**:123-129. DOI: 10.1016/j.molstruc.2012.04.045

[47] Bedia A, Bedia F, Aillerie A. Optical, electrical and structural properties of nano-pyramidal ZnO films grown on glass substrate by spray pyrolysis technique. Optical Materials. 2012;**36**:1123-1130. DOI: 10.1016/j.optmat.2014.02.012

[48] Karber E, Raadik T, Dedova T. Photoluminescence of spray pyrolysis deposited ZnO nanorods. Nanoscale Research Letters. 2011;**6**:359. DOI: 10.1186/1556-276X-6-359

[49] Li Z, Hu Z, Liu F. Lateral growth and optical properties of ZnO microcrystal on sapphire substrate. Optical Materials. 2012;**34**:1908-1912. DOI: 10.1016/j.optmat.2012.05.033

[50] Khan Z, Khan M, Zulfequar M. Optical and structural properties of ZnO thin films fabricated by sol-gel method. Materials Sciences and Applications. 2011;**52**:340-345. DOI: 10.4236/msa.2011.25044

[51] Kishore Kumar Y, Uday Bhaskar P, Suresh Babu G. Effect of copper salt and thiourea concentrations on the formation of Cu_2ZnSnS_4 thin films by spray pyrolysis. Physica Status Solidi. 2010;**207**:149-156. DOI: 10.1002/pssa.200925194

[52] Fernandes P, Salome P, da Cunha A. Study of polycrystalline Cu_2ZnSnS_4 films by Raman scattering. Journal of Alloys and Compounds. 2011;**509**:7600-7606. DOI: 10.1016/j.jallcom.2011.04.097

[53] Patel M, Mukhopadhyay I, Ray A. Structural, optical and electrical properties of spray-deposited CZTS thin films under a non-equilibrium growth condition. Journal of Physics D. 2012;**45**:445103. DOI: 10.1088/0022-3727/45/44/445103

[54] Fernandes P, Salome P, da Cunha A. Growth and Raman scattering characterization of Cu_2ZnSnS_4 thin films. Thin Solid Films. 2009;**517**:2519-2523. DOI: 10.1016/j.tsf.2008.11.031

[55] Fontane X, Calvo-Barrio L, Izquierdo-Roca V. In-depth resolved Raman scattering analysis for the identification of secondary phases: Characterization of Cu_2ZnSnS_4 layers for solar cell applications. Applied Physics Letters. 2011;**98**:181905. DOI: 10.1063/1.3587614

[56] Dumcenco D, Huang Y. The vibrational properties study of kesterite Cu_2ZnSnS_4 single crystals by using polarization dependent Raman spectroscopy. Optical Materials. 2013;**35**:419-425. DOI: 10.1016/j.optmat.2012.09.031

[57] Kodigala S. Thin film solar cells from earth abundant materials: Growth and characterization of $Cu_2(Zn,Sn)(S,Se)_4$ thin films and their solar cells. London: Newnes; 2013. 190p. ISBN: 9780123944290

Chapter 5

E-ALD: Tailoring the Optoeletronic Properties of Metal Chalcogenides on Ag Single Crystals

Emanuele Salvietti, Andrea Giaccherini,
Filippo Gambinossi, Maria Luisa Foresti,
Maurizio Passaponti, Francesco Di Benedetto and
Massimo Innocenti

Additional information is available at the end of the chapter

http://dx.doi.org/10.5772/intechopen.71014

Abstract

Technological development in nanoelectronics and solar energy devices demands nanostructured surfaces with controlled geometries and composition. Electrochemical atomic layer deposition (E-ALD) is recognized as a valid alternative to vacuum and chemical bath depositions in terms of growth control, quality and performance of semiconducting systems, such as single 2D semiconductors and multilayered materials. This chapter is specific to the E-ALD of metal chalcogenides on Ag single crystals and highlights the electrochemistry for the layer-by-layer deposition of thin films through surface limited reactions (SLRs). Also discussed herein is the theoretical framework of the under potential deposition (UPD), whose thermodynamic treatment open questions to the correct interpretation of the experimental data. Careful design of the E-ALD process allows fine control over both thickness and composition of the deposited layers, thus tailoring the optoelectronic properties of semiconductor compounds. Specifically, the possibility to tune the band gap by varying either the number of deposition cycles or the growth sequence of ternary compounds paves the way toward the formation of advanced photovoltaic materials.

Keywords: E-ALD, Ag(111), thin films, metal chalcogenides, UPD, photovoltaics

1. Introduction

Improving the efficiency of the deposition techniques of compound semiconductors could be crucial for the future sustainable micro and nano-electronic industry. A common challenge when reviewing current technologies is the lack of reliable compositional control as well as conformal coating of complex geometries. Electrochemical deposition techniques are a low-cost alternative to vacuum evaporation and chemical bath deposition for the direct fabrication of thin films from molecular precursors. Among electrodeposition processes, electrochemical atomic layer deposition (E-ALD) has the advantage to fabricate semiconductors one atomic layer at a time, thus requiring very low energy consumption, diluted solutions, and operating at standard environmental conditions. Fine control of the thin film growth is achieved through the surface limited reaction (SLR), usually referred as under potential deposition (UPD), of the ionic precursors with the electrode substrate or a preceding layer. When the thin film growth can be rigorously considered epitaxial, the E-ALD method is referred as electrochemical atomic layer epitaxy (ECALE) [1]. E-ALD is recognized as a very effective method for the electrodeposition of ultra-thin films of semiconducting materials. In recent years, thin films of binary [2–5] and ternary semiconductors [6–9] were successfully obtained on silver electrodes. **Figure 1** illustrates the steps involved in the E-ALD process of a ternary compound. The chosen sequence for the alternate deposition of different elements dictates both the structure and the stoichiometry of the resulting semiconductor compound.

This chapter highlights the current progress in E-ALD of binary and ternary chalcogenides on Ag(111) from an electrochemical perspective.

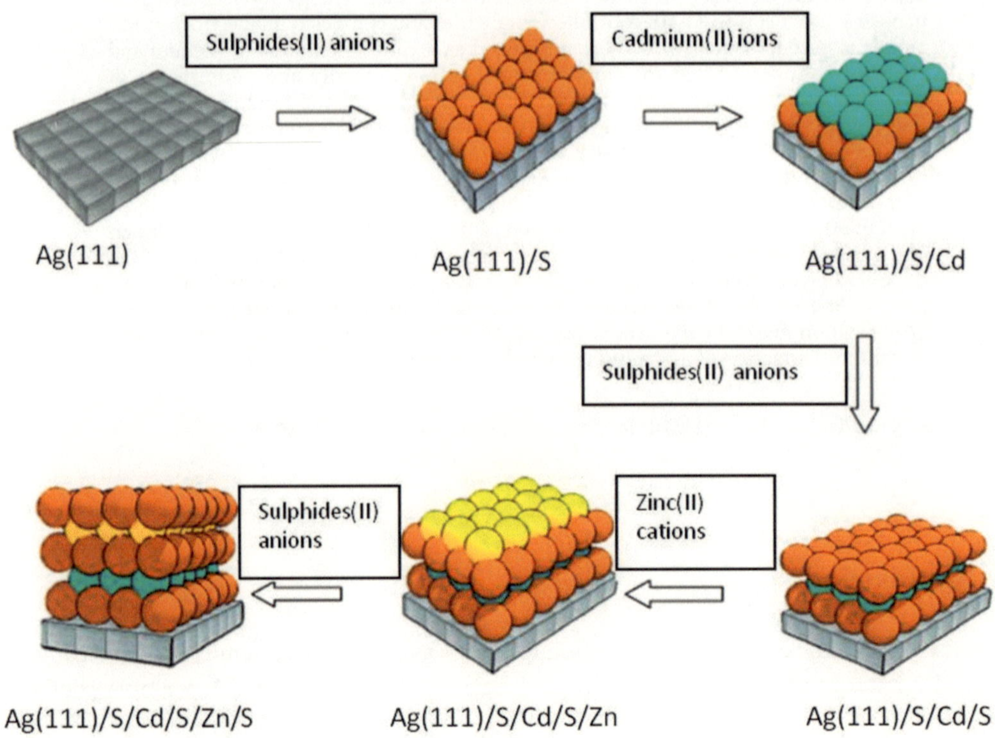

Figure 1. Schematic of the E-ALD growth of a ternary semiconductor. The solutions used in each step are indicated in the boxes below.

2. Thermodynamics of the UPD process

For the sake of simplicity and clarity, we will develop the thermodynamics of an epitaxial E-ALD process (ECALE) with the steps constituted by UPD. As a reference we consider the ECALE growth of CdS on Ag(111). From a purely chemical perspective, ECALE exploits UPD to grow the semiconductor compounds on a well-behaving surface. For instance, the oxidative deposition of an atomic layer of sulfur atoms on top of Ag(111) surface is:

$$Ag(111) - S_{ad} + 2e^{-} \leftrightarrows Ag(111) + S^{2-} \tag{1}$$

where S_{ad} refers to the sulfur adlayer. The UPD provides a surface limited deposition process, which is characterized by a potential more cathodic than the bulk deposition. The latter is defined by means of the Nernst equation:

$$E_{S^{2-}/S} = E^{0}_{S^{2-}/S} + \frac{RT}{2F} ln\left(\frac{a_S}{a_{S^{2-}}}\right) \tag{2}$$

Considering the experimental observation of the potential shift provided by a UPD process, a heuristic extension of the Nernst equation can be proposed:

$$E_{S^{2-}/S_{Ad}} = E^{0}_{S^{2-}/S} + \frac{RT}{2F} ln\left[\frac{a_{S_{ad}}(\theta)}{a_{S^{2-}}}\right] \tag{3}$$

where $a_{S_{ad}}(\theta)$ is the activity of S atoms adsorbed on Ag(111) as a function of the coverage θ. Rearranging Eqs. (2) and (3) we obtain the underpotential shift, ΔE_{UPD} [10]:

$$\Delta E_{UPD} = E_{S^{2-}/S_{Ad}} - E_{S^{2-}/S} = \frac{RT}{2F} ln\left[\frac{a_{S_{ad}}(\theta)}{a_S}\right] \tag{4}$$

which is, incorrectly, independent from the activity of the sulfide anions. This heuristic approach, though enabling an intuitive description of the UPD process, does not take into account other terms such the local defects of the electrode surface, and the mutual interactions between solvent, surfaces and anions.

A more accurate description of the UPD thermodynamics can be formulated in the framework of the ideal polarized electrode. Following this approach, the substrate (i.e. Ag(111)) on which UPD takes place is in contact with the electrolyte solution (sol) containing S^{2-}, Na^{+} and X^{-} ions and solvent (H_2O). From a theoretical standpoint, the electrolyte solution can be considered in contact with a reference electrode reversible with respect to a general anion X^{-}. The phases relevant for the present analysis are the Ag(111) surface, the solution (sol), and the interphase (IP) between the substrate and sol. Each phase, j, is described thermodynamically by the Gibbs energy, G^{j}, which is a function of the variables T^{j} (temperature), P^{j} (pressure), A^{IP} (interfacial area) and N^{j}_{i} (amount of matter of the ith species):

$$G^{j} = G^{j}\left(T^{j}, P^{j}, A^{IP}, N^{j}_{i}\right) \tag{5}$$

The total derivative of the Gibbs energy gives [10]:

$$dG = \sum_j dG^j = \gamma^{IP} dA^{IP} + \sum_{j,i} S^j dT^j - V^j dP^j + \mu^j dN_i^j \tag{6}$$

where γ^{IP} is the surface tension of the interphase, S^j the entropy, V^j the volume and μ^j the chemical potential.

Let us now consider a solution of Na_2S in NaOH buffer solution (i.e. pH = 13) at constant T^j, P^j. The main charge transfer equilibrium is:

$$S^{2-}_{sol} \leftrightarrows S_{el} \tag{7}$$

where *sol* refers to the liquid phase and *el* refers to the electrode. The corresponding electrocapillary equation is [11, 12]:

$$-d\gamma^{IP} = \Gamma^{IP}_{S^{2-}} d\mu^{sol}_{S^{2-}} + \Gamma^{IP}_{OH^-} d\mu^{sol}_{OH^-} + \Gamma^{IP}_{S} d\mu^{el}_{S} - q_{ion} dE \tag{8}$$

where E is the potential referred to the reference electrode and it is defined by means of the chemical potential $\Gamma^{IP}_{Na^+} d\mu^{sol}_{Na^+}$. According to this latter definition, q_{ion} is the surface excess of the electric charge density:

$$q_{ion} = \left(\frac{\partial \gamma^{IP}}{\partial E}\right)_{\left(\mu^{sol}_{S^{2-}}, \mu^{sol}_{OH^-}, \mu^{el}_{S}\right)} \tag{9}$$

also known as Lippmann equation. $\Gamma^{IP}_{OH^-}$, $\Gamma^{IP}_{S^{2-}}$ and Γ^{IP}_{S} are the relative excess of Na^+ and S^{2-} with respect to H_2O (solvent):

$$\Gamma^{IP}_{S^{2-}} = \frac{1}{A^{IP}} \left[N^{IP}_{S^{2-}} - \frac{N^{sol}_{S^{2-}} N^{IP}_{H_2O}}{N^{sol}_{H_2O}} \right] \tag{10}$$

It is important to notice that Eq. (10), in the Gibbs theoretical framework, is not a mere change of variable. In fact, the amounts of matter (N^{IP}_i) depend on the arbitrary definition of the interface while the relative excess of matter (Γ^{IP}_i) is independent of the area and the thickness of the interface. Moreover, the contribution of the formation of the S adlayer is described by the term $d\gamma^{IP}$. In this context, using the following well-known equation for the chemical potential in the case of complete dissociation:

$$\mu^{sol}_{S^{2-}} = \mu^{0,sol}_{S^{2-}} + RT \ln a_{S^{2-}} \tag{11}$$

$$\mu^{sol}_{OH^-} = \mu^{0,sol}_{OH^-} + RT \ln a_{OH^-} \tag{12}$$

$$\mu^{el}_{S} = \mu^{el}_{S} + RT \ln \theta_S \tag{13}$$

where a is the activity in the liquid phase and θ is the fractional amount in the solid phase. In this framework the relationship between activities and surface tension is well represented by the following:

$$\Gamma_{S^{2-}}^{IP} = \frac{1}{RT}\left(\frac{\partial \gamma^{IP}}{\partial \ln a_{S^{2-}}}\right)_{\left(a_{OH^-}^{sol},\theta_S^{el},q_{ion}\right)} \tag{14}$$

$$\Gamma_{OH^-}^{IP} = \frac{1}{RT}\left(\frac{\partial \gamma^{IP}}{\partial \ln a_{OH^-}}\right)_{\left(a_{S^{2-}}^{sol},\theta_S^{el},q_{ion}\right)} \tag{15}$$

$$\Gamma_{S}^{IP} = \frac{1}{RT}\left(\frac{\partial \gamma^{IP}}{\partial \ln \theta_S}\right)_{\left(a_{S^{2-}}^{sol},a_{OH^-}^{sol},q_{ion}\right)} \tag{16}$$

$$E = \left(\frac{\partial \gamma^{IP}}{\partial q_{ion}}\right)_{\left(a_{s^{2-}}^{sol},a_{OH^-}^{sol},\theta_S^{el}\right)} \tag{17}$$

Hence, the surface tension depends on the activity of the chemical species:

$$\gamma = \gamma_{\left(a_{S^{2-}}^{sol},a_{OH^-}^{sol},\theta_S^{el},q_{ion}\right)} \tag{18}$$

Therefore, upon the integration of Eq. (8), an extension of the Nernst equation can be obtained substituting Eqs. (11)–(13) and (18):

$$E = E^0 + \frac{\Gamma_{S^{2-}}^{IP} RT}{q_{ion}} \ln a_{S^{2-}} + \frac{\Gamma_{OH^-}^{IP} RT}{q_{ion}} \ln a_{OH^-} + \frac{\Gamma_{S}^{IP} RT}{q_{ion}} \ln \theta_S + \frac{1}{q_{ion}}\left[\gamma_{\left(a_{S^{2-}}^{sol},a_{OH^-}^{sol},\theta_S^{el},q_{ion}\right)} - \gamma^0\right] \tag{19}$$

where

- E^0 is the standard potential of the electrode with respect to the reference electrode,
- $\frac{\Gamma_{S^{2-}}^{IP} RT}{q_{ion}} \ln a_{S^{2-}}$ takes into account the effect of the activity,
- $\frac{\Gamma_{OH^-}^{IP} RT}{q_{ion}} \ln a_{OH^-}$ takes into account the effect of the pH, constant in a buffer solution,
- $\frac{\Gamma_{S}^{IP} RT}{q_{ion}} \ln \theta_S$ takes into account the effect of the surface coverage: $\theta_S = 1$ for bulk phases, θ_S the fractional amount for compounds and the coverage of adlayers,
- $\frac{1}{q_{ion}}\left[\gamma\left(a_{S^{2-}}^{sol}, a_{OH^-}^{sol}, \theta_S^{el}, q_{ion}\right) - \gamma^0\right]$ takes into account the interaction with Ag(111),

Eventually, we should state that Eq. (19), in principle, takes into account all the effects neglected by Eq. (3), thus describing the equilibrium of a system defined by three components, as predicted by the phase rule. Experimental observations suggest that the potential of the UPD process of sulfur depend on the concentration of sulfur atoms ($a_{S^{2-}}$), the pH (a_{OH^-}) and the coverage of the adlayer (θ_S).

3. E-ALD of metal chalcogenides

E-ALD of metal chalcogenides involves sequential SLR of metal and chalcogenides ions. Except for few cases, the choice of chalcogenides as first layer is preferred for its higher affinity for silver and to avoid alloy formation with the electrode substrate [2]. In the E-ALD methodology a monolayer of the compound is obtained by alternating underpotential deposition of the metallic element with the underpotential deposition of the non-metallic element in a cycle, so the thickness of deposited film is a function of the numbers of deposition cycles. **Figure 2a** shows the cyclic voltammogramm of 0.5 mM Na_2SeO_3 on Ag(111) and of $ZnSO_4$ on Se-covered Ag(111). The cathode peak C_{Se} at −0.95 V corresponds to the bulk reduction of the Se. At more negative potentials hydrogen evolution prevents the detection of the Se_{ad} reduction. Zn adlayers deposited on a Se-covered Ag(111) show a CV profile shifted to more positive potential with a reduction peak C_{Zn} at −0.65 V. To prevent Se dissolution, the operating potential for the surface limited deposition of Zn on Ag/Se should be more positive than C_{Se}. Anodic stripping analysis as a function of the applied potential allow to choosing the optimal conditions for the UPD process, which in this case is in the range $-0.95 < E < -0.8$ V, where a plateau at $Q = 60$ μC cm^{-2} is observed (see **Figure 2b**). Alternate depositions of both chalcogenide and metal can be repeated as many times as necessary to form a compound with the desired thickness and composition. **Figure 2c** shows the charges involved in the layer-by-layer growth of Zn and Se in a 1:1 ratio.

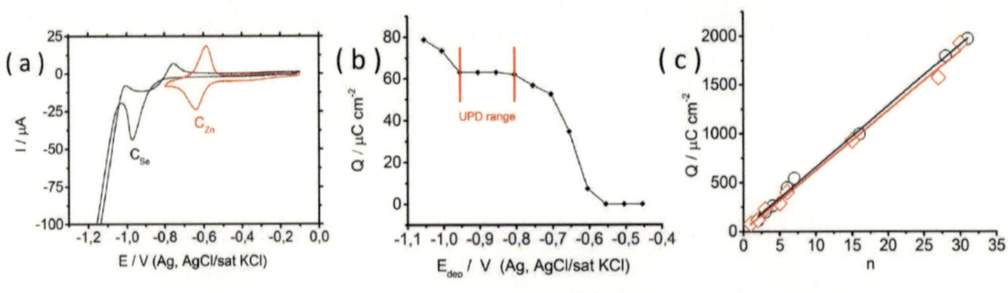

Figure 2. (a) Cyclic voltammograms of (—) Se on Ag(111) and (—) Zn on Se-covered Ag(111) in ammonia buffer solutions (pH 9.2); (b) charges involved in the stripping of Zn, underpotentially deposited on Ag(111)/Se_{ad}, as a function of the deposition potential; (c) charges involved in the striping of (○) Se and (□) Zn as a function of the number of E-ALD cycles. The solid lines represent the linear fit to the data.

4. Flow cell apparatus and experimental conditions

E-ALD thin films of metal chalcogenides were grown on silver single crystal disks using an automated deposition apparatus consisting of Pyrex solution reservoirs, solenoid valves, a distribution valve and a flow-cell [2]. The solutions contained in the Pyrex reservoirs are previously degassed and then constantly kept under a nitrogen pressure $p(N_2) = 0.3$ atm. **Figure 3a** shows the distribution valve with seven solution inlets, in the top, and one solution

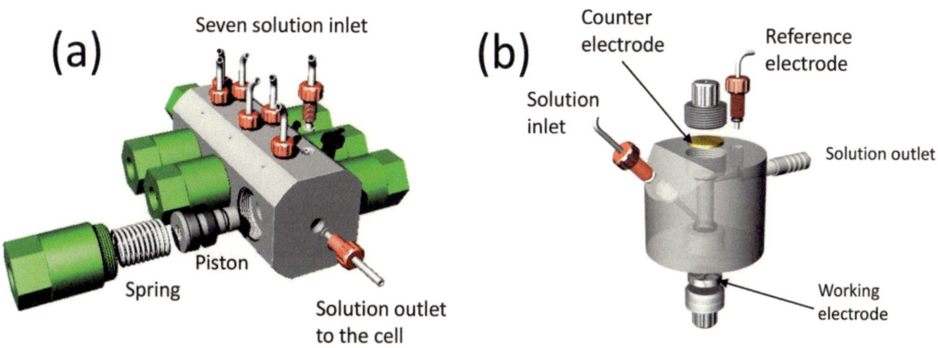

Figure 3. (a) Distribution valve and (b) electrochemical cell.

outlet, in the front. The solution is pushed to the distribution valve by the N_2 overpressure. The inlet of the solution is blocked by a piston of the distribution valve, which is tightly held in place by a spring. The piston can be raised by opening the solenoid valve and sending compressed air at 6 atm, which is higher than the pressure exerted by the spring. The piston is raised enough to allow the solution to flow into the cell. Different solutions are pushed to the cell following the desired sequence by acting on the corresponding solenoid valves. The nitrogen pressure exerted on the solutions determines a flow rate of about 1 mL s^{-1}. All operations are carried out under computer control.

The electrochemical cell (**Figure 3b**) is a Teflon cylinder with a height of 40 mm, an internal diameter of 10 mm and external diameter of 50 mm. The working electrode at the bottom of the cylinder and the counter electrode on top delimit the electrochemical cell volume (1.6 mL). The solution inlet and outlet are placed on the side walls of the cylinder; for hydrodynamic reasons, the inlet is inclined. The working electrode was silver disks cut according to the Bridgman technique [13]. The counter electrode was a gold foil, and the reference electrode was an Ag/AgCl (saturated KCl) placed in the outlet tubing. Leakage is avoided by pressing both the working and the counter electrode against a suitable Viton® O-ring. All potentials reported in the paper are quoted with respect to the Ag/AgCl (saturated KCl) reference electrode.

5. E-ALD of binary M_xS_y semiconductors on Ag(111)

E-ALD technique has been successfully used to fabricate ultrathin films of metal sulfides on silver electrodes by alternating the underpotential deposition of metal and sulfur. These compounds include cadmium sulfide (CdS) [2, 14, 15], zinc sulfide (ZnS) [2], nickel sulfide (NiS) [4], lead sulfide (PbS) [16], copper sulfides (Cu_xS) [17, 18] and tin sulfides (Sn_xS_y) [19]. A typical E-ALD cycle includes the underpotential deposition of sulfur followed by the surface limited reaction (SLR) of metal on S-covered Ag. The UPD of sulfur on crystalline and polycrystalline silver have been extensively investigated in the past [20–22]. Electrochemical measurements on Ag(110), Ag(100) and Ag(111) show that sulfur UPD deposition processes differ significantly on the three silver facets and in situ STM experiments have evidenced the presence of differently ordered sulfur structures depending on substrate orientation. The formation of the first layer of S on Ag(100) and Ag(110) occurs at a potential of $E = -0.8$ V in pH 13 solutions, whereas on Ag (111) it was obtained at $E = -0.68$ V in ammonia buffer (pH 9.6). Proceeding toward more positive potentials in the presence of sulfide ions resulted in bulk sulfur deposition. Cyclic voltammogramms performed in Na_2S solutions revealed two distinct behaviors. While on Ag (100) only two broad anodic peaks at −1.32 and −1.15 V were observed, cycling the potential on the other two faces resulted in a more complex behavior, with a sharp anodic peak occurring at $E = -1.06$ and $E = -0.78$ on Ag(110) and Ag(111), respectively. The charges associated with the UPD deposition of S in the first layer, as calculated from chronocoulometry experiments, were 137μC cm^{-2} for Ag(100), 163μC cm^{-2} for Ag(110) and 189μC cm^{-2} for Ag(111). Although the growth of M_xS_y on silver electrodes follows an oxidative-reductive behavior in all cases, SLR of metal layers depends on the semiconductor type and solution conditions.

5.1. E-ALD of PbS and NiS

Lead sulfide (PbS) and nickel sulfide (NiS) are binary semiconductors that have received considerable attention for a variety of applications, such as detectors [23], sensing materials [24] and solar cells [25]. A complete electrochemical study of PbS multilayers has been reported by Fernandes et al. [16]. The Pb_{ad} cannot be formed on S-covered silver electrodes because of partial redissolution of sulfur. So the first layer is the Pb_{UPD} deposited on Ag(111), this was obtained from 5.0 mM $Pb(NO_3)_2$ solutions in acetic buffer at pH 5 by scanning the potential from −0.2 to −0.45 V. Two well-defined peaks were observed at −0.35 and −0.29 V for the deposition and the dissolution of Pb monolayer, respectively. Next, the underpotential deposition of S on Pb-covered Ag(111) was obtained by scanning the potential from −1.0 to −0.70 V in 2.5 mM Na_2S solutions in ammonia buffer. The constancy of the anodic stripping of Pb deposited at −0.45 V and at different accumulation times ensures the process is surface limited, giving rise to a charge of 332 μC cm^{-2} for the first layer. Successive sulfur and lead layers resulted in a linear growth with an average charge per cycle of approximately 83 μC cm^{-2}. Morphological analysis by ex-situ AFM measurements revealed the deposits consisted of homogeneous films of PbS small clusters.

Differently from lead, underpotential deposition of nickel on bare Ag(111) is not possible due to weak adhesion with the electrode substrate and competing surface phase transformations [26]. On the contrary nickel presents a well-defined surface limited reaction on S-covered Ag (111), showing cathodic and anodic peaks at −0.52 and −0.22 V, respectively. UPD layers of nickel were obtained from $NiCl_2$ in boric acid solutions (pH 6.5) at $E = -0.6$ V. The amount of sulfur, as determined by separate anodic stripping experiments, was found to increase linearly with the number of the deposition cycles. Conversely, the stripping of nickel was less precise due to the formation of oxide and hydroxide films with the increase in layer number [27]. Despite these limitations the charges of both Ni and S showed a quasi-linear layer-by-layer growth with slopes of 53 μC cm^{-2} for Ni and 58 μC cm^{-2} for S. Morphological analysis by AFM indicated a decrease of average roughness with aging, thus suggesting the formation of a passivation layer with time, which was later confirmed by XPS analysis.

5.2. E-ALD of CdS, ZnS, Sn_xS_y and Cu_xS

Cadmium, tin, zinc, and copper are metals that form strong interactions with sulfur ions, thus generating binary semiconductors with peculiar transport and electrical properties. Cadmium and zinc sulfides have been among the first binary semiconductors to be deposited by means of E-ALD due to favorable electrochemical characteristics.

UPD deposition of CdS has been investigated on silver single crystals exhibiting different orientations, that is, Ag(111), Ag(110) and Ag(111) [14, 15].The well-defined structures of S_{UPD} formed on silver electrodes in Na_2S ammonia buffer solutions drive the epitaxial deposition of the subsequent Cd layer. After washing out the sulfur ions in excess, UPD of cadmium is performed by holding the potential at −0.68 V on S-Ag(111) and at E = −0.6 V on the other two faces. The procedure used to obtain further alternate S and Cd layers is identical with that used for the Ag/S/Cd structure. The average values of the charge deposited in each E-ALD cycle were 103 μC cm^{-2} for Ag(100), 85 μC cm^{-2} for Ag(110) and 70 μC cm^{-2} for Ag(111). Ex situ XPS analysis confirmed the presence of cadmium and sulfur in a 1:1 stoichiometric ratio [28, 29].

Similarly to cadmium sulfide, E-ALD of ZnS thin film were grown on silver electrodes by first depositing sulfur at E = −0.70 V from Na_2S solution, and then injecting $ZnSO_4$ solution while keeping the electrode at the same potential to deposit Zn underpotentially. Plots of the charges for Zn and S measured in the stripping of ZnS deposits were linear, with a slope of 67 μC cm^{-2} for Zn and 75 μC cm^{-2} for S [2].

Differently from CdS and ZnS, the other two metal sulfides (Cu_xS and Sn_xS_y) exist in different stoichiometric ratios and morphologies; their interest in E-ALD growth relies on the tunable transport and electronic properties by changes in composition, x and y [30, 31].

The surface limited reaction of Sn_xS_y has been thoroughly studied by Innocenti et al. [19] both on bare and on S-covered silver substrates. Electrodeposition of tin on bare silver showed two anodic peaks, E_{c1} = −0.70 V and E_{c2} = −0.48 V, the latter ascribed to the formation of Sn(IV) hydroxides. Differently, on S-covered Ag(111) the reduction peak was seen at lower potentials, E_c = −0.61 V, thus suggesting true underpotential deposition mechanism. Independently from the deposition time, the charge involved in the oxidative process remained nearly constant, thus confirming a surface limited process.

Thin films of copper sulfides were fabricated on silver substrates through E-ALD. Although surface limited the layer-by-layer growth of Cu_xS was found not to be a true UPD process like for the other metal sulfides. As reported by Innocenti et al. [17] the electrochemistry of copper on S-covered Ag(111) is quite complex. Cyclic voltammograms as obtained by sweeping the potential between −0.05 and −0.55 V in 1 mM Cu(II) solutions in ammonia buffer revealed the presence of two cathodic peaks, E_1 = −0.39 V and E_2 = −0.42 V. While the latter was easily associated to the bulk reduction of Cu(II) to Cu(0), the nature of the first cathodic peak is still under debate, except for the fact that it precedes bulk deposition. More-in-depth electrochemical analysis suggests the process is surface limited in the range −0.3 to −0.38 V and it involves the formation of Cu(0) through the two-step reduction of cupric tetra-amino complex and the subsequent conproportionation reaction:

$$Cu(NH_3)_4^{2+} + e^- \rightarrow Cu(NH_3)_2^+ + 2NH_3 \quad (20)$$

$$Cu(NH_3)_2^+ + e^- \rightarrow Cu(0) + 2NH_3 \quad (21)$$

$$Cu(NH_3)_4^{2+} + Cu(0) \rightarrow 2Cu(NH_3)_2^+ \quad (22)$$

The amount of Cu deposited in a given number of cycles was determined by measuring the charge involved in the anodic stripping. The authors found the charges were linearly increasing with the number of deposition cycles with a slope of 44 μC cm^{-2}. XPS results confirmed the valence states of copper and sulfur as Cu(I) and S(−II), respectively, although a possible fraction of S(−I) in the form of disulfide anion was not excluded. The experimental Cu/S ratio observed in the XPS characterization was later attributed to the covellite phase, where positive holes allow Cu ions to be stabilized in their monovalent state [32]. The morphological analysis by AFM was able to evidence the low roughness values of the deposits, thus confirming the high homogeneity and good quality of the thin film obtained.

5.3. E-ALD of ternary $M_xN_{1-x}S$

The interest in cadmium, zinc, tin, and copper sulfides has increased in the last 10 years due to the possibility to fabricate ternary semiconductor compounds, thus allowing fine control over the band gap energy of solar cell devices [32–34]. E-ALD method has been successfully employed to grow ternary materials such as $Cd_xZn_{1-x}S$ [35, 36], $Cu_xZn_{1-x}S$ [5] and $Cu_xSn_yS_z$ [33]. These semiconductors were prepared by sequential deposition of the corresponding binaries; for instance, alternate deposition of Cu_xS and Sn_xS_y was carried out to form $Cu_xSn_yS_z$. Because of the large variety of the possible (x:y:z) combinations research on E-ALD of multinary kesterite group thin films, although challenging, is quite promising for the development of non-linear electro-optic devices and photovoltaic cells. Depending on the adopted sequence profile only certain combinations were attainable, thus limiting the possible metal-to-metal stoichiometries. The electrochemical behavior of $Cu_xSn_yS_z$, $Cu_xZn_{1-x}S$ and $Cd_xZn_{1-x}S$ thin films is separately discussed below.

Di Benedetto et al. [33] investigated the electrodeposition of $Cu_xSn_yS_z$ thin films exploiting different sequences of E-ALD cycles, that is $Ag/S[(Cu/S)_k/(Sn/S)_j]_n$ with $k = 1$, $j = 1, 2$ and $1 < n < 60$. Surface limited deposition of Cu and Sn occurred, respectively, at $E = -0.37$ V and $E = -0.68$ V in ammonia buffer containing EDTA. As already reported for binary sulfides S_{UPD} layers on Ag(111) and on metal were obtained by keeping the potential at $E = -0.68$ V in Na_2S ammonia solutions. Stripping analysis of the ternary sulfides yielded a large and well-defined peak centered at $E = -0.22$ V (Cu stripping), preceded by a broader peak at −0.43 V (Sn stripping). Charges involved in the stripping of both metals (Sn + Cu) and S allowed defining the effective layer-by-layer formation of a ternary compound with a slope of 42 μC cm^{-2}, which is very close to the value obtained from the stripping of the binary CuS compound [19]. The chemical composition of $Ag/S[(Cu/S)_k/(Sn/S)_j]_n$ deposits were analyzed by means of SEM, XPS and TOF-SIMS [37]; the ex-situ characterizations have highlighted the nominal stoichiometry is not respected leading to Sn/Cu ratio equal to 1/13 and 1/9 for $j = 1$, and 2, respectively. Independently from Sn/Cu ratio the morphology of the growing films, as revealed by AFM measurements, were found to be homogeneous and similar to the bare Ag(111). Conversely, the band gap values, carried out by diffuse reflectance spectroscopy, decreased with the increase in thickness and Cu/Sn ratio ranging from 2.12 for $n = 60$, $j = 2$–2.43 eV for $n = 20$, $j = 1$.

As for the Cu-Sn-S system the experimental conditions to grow ternary cadmium and zinc sulfides on Ag(111) with the E-ALD were by alternating the underpotential deposition of the corresponding binaries (CdS and ZnS) [36]. The potential chosen for the deposition of S and Zn was −0.65 V, whereas different series of experiments were carried out depositing Cd in correspondence to the first (−0.5 V) or to the second UPD (−0.65 V). The authors investigated the electrodeposition of $Cd_xZn_{1-x}S$ thin films exploiting different sequences of E-ALD cycles, that is, $Ag/S[(Cd/S)_k/(Zn/S)_j]_n$ with $k = 1$, $j = 1, 2, 5$ and $1 < n < 20$. The charge involved in the stripping increases linearly with the number of deposition cycles. Yet, the slope of the plot decreases while increasing the number of ZnS cycles per CdS cycles, that is, 89, 67 and 52 $\mu C\ cm^{-2}$ for $j = 1, 2$ and 5, which reflect the higher percentage of Zn in the deposit. The chemical composition of $Cd_xZn_{1-x}S$ thin films, as analyzed by XPS have highlighted the nominal stoichiometry is not respected leading to Zn/Cu ratio equal to 1/3, 1/2 and 2 for $j = 1, 2$ and 5, respectively. However, regardless of the stoichiometry of the obtained ternary compound, the charge involved in cathodic stripping was equal to the charge involved in the anodic one, thus indicating the right 1:1 stoichiometric ratio [35].

The low contribution of Zn in $Cd_xZn_{1-x}S$ compounds was also found to occur in the Cu-Zn-S system. Ternary Cu_xZn_yS compounds were obtained through the E-ALD method by alternating deposition of Cu_xS and ZnS layers. Innocenti et al. [8, 38] studied the electrochemical and compositional behavior of $Cd_xZn_{1-x}S$ thin films by applying the general sequence $Ag/S/[(Cu/S)_j/(Zn/S)_k)]_n$ with $j = 1$, $k = 1, 5, 9$, and $1 < n < 60$. As for other binary and ternary compounds UPD sulfur layers were obtained by keeping the electrode potential at E = −0.68 V for either on bare Ag(111) or on Ag(111) already covered by a metal layer. Deposition of Cu and Zn occurred in ammonia buffer at −0.37 and −0.85 V, respectively. Stripping analysis revealed the charge involved in (Cu + Zn) deposition followed a non-linear increase with the number of cycles. Moreover, samples containing higher percentage of Zn had lower slope, thus indicating a lower extent of deposition in each cycle; a Cu/Zn ratio of about 6 was found for $j = 1$, $k = 1$, and $n = 40$. From the plot of charges against the number of deposition cycles (n) we expect a ZnS:CuS equimolar ratio for n = 13. Zn deficiency upon increasing the cycles is due to partial redissolution of zinc during the E-ALD of copper, thus causing a rearrangement in the adlayers. Independently from the Cu/Zn ratio, ex-situ investigations highlighted at least two prevailing morphologies, the first one homogeneously covering the Ag(111) surface, and the second one consisting of random network of nanowires of variable length [8].

6. E-ALD of binary M_xSe_y semiconductors on Ag(111)

The E-ALD of metal selenides (M_xSe_y) on monocrystalline surfaces usually starts with the deposition of a selenium monolayer by using selenite, Se(IV), as a precursor salt. Differently from sulfur deposition the oxidative UPD of Se is not allowed due to the low stability of selenite solutions. According to the work by Rajeshwar et al. [39] the electrochemistry of Se (IV) is quite complex. The direct reduction of Se(IV) to Se(0) is attributed to the formation of the electroinactive gray Se, which forms a stable deposit:

$$H_2SeO_3 + 4H^+ + 4e^- \rightarrow Se + 3\,H_2O \tag{23}$$

Instead, the electroactive form of Se(0), usually referred as red Se, is obtained through a first reduction of Se(IV) to Se(−II) followed by a comproportionation reaction [40]:

$$H_2SeO_3 + 6H^+ + 6e^- \rightarrow H_2Se + 3\,H_2O \tag{24}$$

$$H_2SeO_3 + H_2Se \rightarrow 3Se + 3\,H_2O \tag{25}$$

In ammonia buffer (pH 9.3), the presence of the electroactive red Se is evident for $E < -0.95$ V through the reduction Se(0) → Se(−II), and around $E = -0.8$ V through the oxidation Se (0) → Se(IV). Adlayers of Se on the electrode surface, Se_{ad}, are formed through a two-step procedure involving the deposition of an excess of Se(0) from Se(IV) solutions, followed by the reduction of bulk Se at sufficiently negative potential ($E \approx -0.95$ V). The reduction must be performed in the absence of Se(IV) to avoid the comproportionation reaction with Se(−II) leading to a massive formation of Se(0). The STM investigation at potential more negative than the bulk reduction peak has shown two distinct structures. The layer at more positive potentials has a ($\sqrt{7} \times \sqrt{7}$) R19.1° structure, with an associated charge of 65 μC cm^{-2}, whereas at more negative potentials it has a ($2\sqrt{7} \times 2\sqrt{7}$) R19.1° structure, with an associated charge of 48 μC cm^{-2} [41].

6.1. E-ALD of CdSe and ZnSe

Cadmiun selenide is a promising material for application as thin film solar cells [42, 43], quantum dots [44, 45] and p-n junctions [46, 47]. ZnSe is an ideal candidate for optoelectronic devices, especially blue laser and blue emitters.

The E-ALD methodology has been successfully used for the growth of CdSe and ZnSe on Ag (111) [48, 49]. For the growth of CdSe, Se_{ad} was deposited from a 0.5 mM Se(IV) in a pH 8.5 ammonia buffer solution for 30s at a fixed potential $E = -0.9$ V; then the solution was replaced with the supporting electrolyte alone, and a potential $E = -0.9$ V was applied for 60s to reduce the bulk Se(0). The reductive underpotential deposition of Cd from a 1 mM $CdSO_4$ in a pH 8.5 ammonia buffer solution on a Se-covered Ag(111) substrate, has 2 peaks. A well-defined UPD peak is observed at $E = -0.41$ V, whereas the beginning of a second UPD peak is observed at $E = -0.69$ V. The second UPD peak cannot be completely recorded, since it overlaps bulk Cd deposition and it can never be isolated from bulk Cd deposition. The optimal conditions for the adlayer formation of Cd on Se-covered Ag(111) are by keeping the electrode at $E = -0.55$ V for 30 s. The plots of the charges for Cd and Se stripping as a function of the number of ECALE cycles are linear, with an average charge per cycle of approximately 75 μC cm^{-2}. The coincidence of the charges associated with each layer of Cd and Se gives a 1:1 stoichiometric ratio between the elements as expected for CdSe.

For the growth of ZnSe, Se_{ad} was deposited from a 0.5 mM Se(IV) in a pH 9.2 ammonia buffer solution for 60 s at potential −0.95 V; then, the solution was replaced with the supporting electrolyte alone, and a potential $E = -0.95$ V was applied for 60 s to reduce the bulk Se(0). The Zn UPD on a Se-covered Ag(111) is obtained at $E = -0.8$ V for 30 s from 1 mM $ZnSO_4$ in a pH 9.2 ammonia buffer solution. Once the deposit was formed, the amount of elements deposited in a given number of cycles was estimated from the charge involved in their stripping. The plots of the charges for Zn and Se stripping as a function of the number of cycles are linear, with a charge per cycle of 61 μC cm^{-2} for Zn and 63 μC cm^{-2} for Se.

6.2. E-ALD of $Cd_xZn_{1-x}Se$

The electrochemical conditions necessary to form CdZnSe deposits by E-ALD on Ag(111) are described in the works by Loglio et al. [36, 50]. The first atomic layer on Ag(111) is obtained by SLR of selenium, as previously described. The UPD deposition on Se-covered Ag(111) from ammonia buffer of Cd would cause Zn redissolution. To shift Cd_{UPD} toward more negative potentials it is necessary to use a stronger complexing agent, that is, 0.1 M pyrophosphate in 0.01 M NaOH, and slightly shifting the Zn UPD toward more positive potentials using acetic buffer at pH 5.0. The experimental procedure to obtain $Cd_xZn_{1-x}Se$ with different x values consists of alternating ZnSe and CdSe ECALE cycles with different deposition sequences. The reductive UPD of Cd on a Se-covered Ag(111) substrate, from a 1 mM Cd(II) solution, was obtained by applying a potential $E = -0.7$ V for 60 s. In an analogous way, ZnSe was obtained by depositing Zn, from a 1 mM Zn(II) solutions, at $E = -0.8$ V. The stripping analysis of the compound obtained with 100 deposition (CdSe + ZnSe) cycles revealed a very high percentage of Cd (74%).

The most effective way to increase the amount of Zn in the deposit consists of depositing more ZnSe cycles per CdSe cycle. The x stoichiometric parameter is a linear function of the ZnSe/CdSe sequence; $x = 0.5$ was obtained with the ZnSe/CdSe sequence equal to 5.

The charge involved in stripping the cations equals the charge involved in stripping the anions, thus confirming the right 1:1 stoichiometric ratio calculated from XPS data. The charge per cycle changes while changing the ZnSe/CdSe sequence, a minimum at $x = 0.5$ is observed, with two symmetrical branches around the minimum. When the percentage of Zn in the compound is approximately equal to that of Cd, the different structures (Zincblend and Wurtzite) are present in a comparable amount, thus suggesting a structural disorder that could be responsible for a reduced amount of deposition.

7. E-ALD of M_xTe_y semiconductors on Ag(111)

The E-ALD of metal tellurides, M_xTe_y, on monocrystalline surfaces starts with the deposition of tellurium. As in the case of selenite, telluride solutions are not stable, so the oxidative UPD is not allowed. Te(IV) reduction occurs following two possible schemes [51, 52]. A stable Te deposit is formed upon direct reduction of Te(IV)

$$H_2TeO_2^+ + 3H^+ + 4e^- \rightarrow Te + 2\,H_2O \tag{26}$$

or through a two-step process involving the reduction of Te(IV) to Te(−II) followed by a comproportionation reaction

$$H_2TeO_2^+ + 5H^+ + 6e^- \rightarrow H_2Te + 2\,H_2O \tag{27}$$

$$H_2TeO_2^+ + 2H_2Te \rightarrow 3Te + 2\,H_2O + H^+ \tag{28}$$

The TeO_2 in ammonia buffer solutions (pH 8.5) on Ag(111) shows a large reduction peak at E = −0.4 V, which is only observed during the first scan from −0.1 to −0.9 V. Integration of the peak yields a charge of about 370 μC cm^{-2}. Due to the high irreversibility of the system, the reoxidation of the underpotentially deposited Te is prevented by silver oxidation, so the Te UPD peak disappears in the successive CV scans. At about −1.1 V only bulk Te(0) reduction occurs. Te UPD layer starts the reduction at potentials more negative than −1.5 V because of higher bonding energy with the silver substrate.

So, the Te_{ad} can be obtained through UPD reduction before bulk deposition or in a two-step process: deposition at a potential of −0.6 V of small excess of bulk Te, followed by the reduction of bulk Te (but not Te_{ad}) at a potential of −1.4 V.

7.1. E-ALD of CdTe

CdTe deposition on Ag(111) was obtained by E-ALD method [53], alternating UPD of tellurium and cadmium. The formation of the first layer of Te on Ag(111) is obtained by a two-step procedure instead of the direct deposition before the bulk reduction, previously described. This choice allows to having the same standard sequence for all cycles, which can be repeated as many times as desired.

The second step of the ECALE cycle is the underpotential deposition of the metallic element on the silver substrate covered by the non-metallic element. UPD of cadmium on Te-covered electrode occurs from 0.5 mM Cd(II) in ammonia buffer solution at a potential of −0.6 V, more negative than on bare Ag(111). Cadmium cyclic voltammograms, obtained from 0.5 mM Cd(II) in an ammonia buffer solution, do not exhibit narrow and sharp peaks as in the case of Te_{ad}. This finding could be attributed to a partial overlap with a second UPD peak, which in turn cannot be easily isolated from the concomitant beginning of bulk deposition. The second Te_{ad} layer cannot be obtained by the direct deposition before the bulk reduction, since underpotential deposition of Te occurs at a more positive potential than Cd stripping. Therefore, the two-step procedure described above has been adopted. The plots of the charges for Cd and Te stripping as a function of the number of cycles are linear, with an average charge per cycle equal to about 175 μC cm^{-2} for Cd and 155 μC cm^{-2} for Te. The ratio Cd/Te determined on the basis of electrochemical measurements is very close to the 1:1 stoichiometric ratio, which is indicative of a compound formation. The linear behavior suggests a layer-by-layer growth.

8. E-ALD of CdS_xSe_{1-x} on Ag(111)

Cadmium chalcogenides such as CdSe and CdS are excellent materials for the development of high efficient and low-cost photovoltaic devices. The small lattice mismatch between CdS and CdSe allows the formation of cadmium sulfoselenides CdS_xSe_{1-x} over a wide range of compositions ($0 < x < 1$), thus covering the visible solar spectrum from $E = 2.44$ eV for $x = 1$ to $E = 1.72$ eV for $x = 0$ [54, 55].

The first ECALE study on ternary CdS_xSe_{1-x} compounds has been reported by Foresti et al. on Ag(111) substrates by alternating the underpotential deposition of CdSe and CdS layers [56]. To obtain CdSe, cadmium was deposited on Se-covered silver substrates from 1 mM Cd(II) solution in ammonia buffer (pH 9.6) by applying a potential $E = -0.5$ V for 60s. The formation of the first layer of Se on Ag(111) were obtained by a two-step procedure as described before [36]. After Cd deposition the cell was rinsed with ammonia buffer and the ECALE cycle was then completed by depositing CdS. The S_{UPD} layer was obtained by applying a potential $E = -0.68$ V for 60s followed by washing with ammonia buffer. Finally, the reductive underpotential deposition of cadmium was attained on a S layer from 1 mM Cd(II) solutions by keeping the electrode at $E = -0.65$ V. The alternated deposition of CdSe and CdS can be repeated as many times as desired to obtained deposits of variable thickness and composition. The authors investigated different sequences of E-ALD cycles, that is $Ag/[(Cd/Se)_j/(Cd/S)_k)]_n$ with $j{:}k = 1{:}1$, $j{:}k = 2{:}3$ and $1 < n < 100$. Independent from the sequence, CdSe deposition appeared to be favored with respect to CdS growth leading to the formation of sulfoselenides with $x = 0.2$ and 0.4 for $j{:}k = 1{:}1$ and 2:3, respectively. The charges involved in the stripping increase linearly with the number of deposition cycles, thus supporting a layer-by-layer growth. Regardless of the stoichiometry of the ternary compound obtained, the two charges are equal, thus confirming the right 1:1 stoichiometric ratio between Cd and (S + Se). Ex situ AFM measurements as a function of composition indicated the roughness decreased while increasing the S percentage, which determines a better deposit. Photoelectrochemical measurements on CdS_xSe_{1-x} ternary compounds revealed a monotonic band gap dependence with x, thus confirming the formation of a single homogeneous phase.

9. Conclusions

The E-ALD is a very inexpensive method for the production of high-crystalline thin film semiconductors. Exploiting Surface Limited Reactions (SLRs) on electrode surfaces allows the layer-by-layer deposition of different atomic elements. The E-ALD has been successfully used to grow ultra-thin films of metal chalcogenides on silver single crystals. The electrochemistry of binary compounds indicates that, with only the exception of the first layer, the charge associated with each layer of either metal or chalcogenide has the same average value. The layer in direct contact with the silver substrate can be regarded as an interface between the metal and the semiconductor electrodeposited on it. Semiconductors grown on Ag(111) by means of E-ALD are characterized by a high crystallinity of the materials and good photoconversion efficiencies. The ability to control the thickness of the deposited layers allows

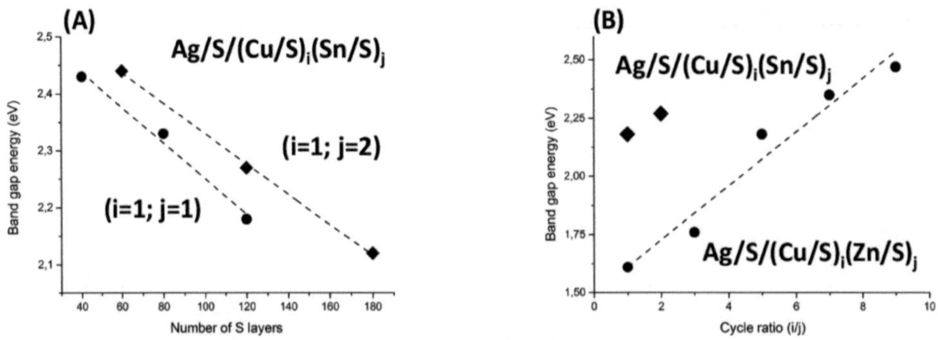

Figure 4. Experimental band gap values (markers) of ternary compounds as a function of (A) thickness at fixed composition and (B) Cu ratio at fixed number of S layers, n = 121. The dotted lines are the linear fit to the data.

fine tuning the semiconductor transition energy by varying the number of deposition cycles (**Figure 4A**). Moreover, careful design of the growth sequence of ternary compounds direct the formation of advanced photovoltaic materials having compositionally controlled band gaps (**Figure 4B**).

Author details

Emanuele Salvietti[1], Andrea Giaccherini[1], Filippo Gambinossi[1], Maria Luisa Foresti[1], Maurizio Passaponti[1], Francesco Di Benedetto[2] and Massimo Innocenti[1]*

*Address all correspondence to: m.innocenti@unifi.it

1 Department of Chemistry, University of Florence, Sesto Fiorentino, Italy

2 Department of Earth Sciences, University of Florence, Florence, Italy

References

[1] Gregory BW, Stickney JL. Electrochemical atomic layer epitaxy (ECALE). Journal of Electroanalytical Chemistry. 1991;**300**:543-561

[2] Innocenti M, Pezzatini G, Forni F, Foresti ML. CdS and ZnS deposition on Ag (111) by electrochemical atomic layer epitaxy. Journal of the Electrochemical Society. 2001;**148**:C357-C362

[3] Innocenti M, Forni F, Pezzatini G, Raiteri R, Loglio F, Foresti ML. Electrochemical behavior of As on silver single crystals and experimental conditions for InAs growth by ECALE. 2001;**514**:75-82

[4] Loglio F, Innocenti M, Jarek A, Caporali S, Pasquini I, Foresti ML. Nickel sulfur thin films deposited by ECALE: Electrochemical, XPS and AFM characterization. Journal of Electroanalytical Chemistry. 2010;**638**(10):15-20

[5] Innocenti M, Becucci L, Bencistà I, Carretti E, Cinotti S, Dei L, Di Benedetto F, Lavacchi A, Marinelli F, Salvietti E, Vizza F, Foresti ML. Electrochemical growth of Cu-Zn sulfides. Journal of Electroanalytical Chemistry. 2013;**710**:17-21

[6] Loglio F, Innocenti M, Pezzatini G, Foresti ML. Ternary cadmium and zinc sulfides and selenides: Electrodeposition by ECALE and electrochemical characterization. Journal of Electroanalytical Chemistry. 2004;**562**(1):17-125

[7] Innocenti M, Becucci L, Bencistà I, Carretti E, Cinotti S, Dei L, Di Benedetto F, Lavacchi A, Marinelli F, Salvietti E, Vizza F, Foresti ML. Ternary cadmium and zinc sulfides and selenides: Electrodeposition by ECALE and electrochemical characterization. Journal of Electroanalytical Chemistry. 2013;**70**:17-21

[8] Innocenti M, Cinotti S, Bencistà I, Carretti E, Becucci L, Di Benedetto F, Lavacchi A, Foresti ML. Electrochemical growth of Cu-Zn sulfides of various stoichiometries. Journal of the Electrochemical Society. 2014;**161**(1):D14-D17

[9] Di Benedetto F, Cinotti S, Guerri A, De Luca A, Lavacchi A, Montegrossi G, Carlà F, Felici R, Innocenti M. Physical characterization of thin films of $Cu_xZn_yS_z$ for photovoltaic applications. ECS Transactions. 2013;**58**(11):59-65

[10] Oviedo OA, Reinaudi L, Garcia SG, Leiva EPM. Underpotential Deposition. Switzerland: Springer; 2015

[11] Parsons R. Comprehensive Treatise of Electrochemistry. Vol. 1. Switzerland: Springer; 1980

[12] Jain S, Willander M, Van Overstraeten R. Compound Semiconductors Strained Layers and Devices. Cham Heidelberg New York: Springer; 2000

[13] Hamelin A. In: Conway BE, White RE, O'M. Bockris J, editors. Modern Aspects of Electrochemistry. Vol. 16. New York: Plenum Press; 1985. p. 1

[14] Foresti ML, Pezzatini G, Cavallini M, Aloisi G, Innocenti M, Guidelli R. Electrochemical atomic layer epitaxy deposition of CdS on Ag(111): An electrochemical and STM investigation. The Journal of Physical Chemistry. B. 1998;**102**(38):7413-7420

[15] Carlà F, Loglio F, Resta A, Felici R, Lastraioli E, Innocenti M, Foresti ML. Electrochemical atomic layer deposition of CdS on Ag single crystals: Effects of substrate orientation on film structure. Journal of Physical Chemistry C. 2014;**118**:6132-6139

[16] Fernandes VC, Salvietti E, Loglio F, Lastraioli E, Innocenti M, Mascaro LH, Foresti ML. Electrodeposition of PbS multilayers on Ag(111) by ECALE. Journal of Applied Electrochemistry. 2009;**39**:2191-2197

[17] Innocenti M, Bencistà I, Bellandi S, Bianchini C, Di Benedetto F, Lavacchi A, Vizza F, Foresti ML. Electrochemical layer by layer growth and characterization of copper sulfur thin films on Ag(111). Electrochimica Acta. 2011;**58**:599-605

[18] Giaccherini A, Cinotti S, Guerri A, Carlà F, Montegrossi G, Vizza F, Lavacchi A, Felici R, Di Benedetto F, Innocenti M. Operando SXRD study of the structure and growth process of Cu2S ultra-thin films. Scientific Reports. 2017;**7**:1615

[19] Innocenti M, Bencista I, Di Benedetto F, Cinotti S, De Luca A, Bellandi S, Lavacchi A, Muniz Miranda M, Vizza F, Marinelli F, Dei L, Carretti E, Zafferoni C, Foresti ML. Underpotential deposition of Sn on S-covered Ag(111). ECS Transactions. 2013;**50**(21):1-7

[20] Hatchett DW, White HS. Electrochemistry of sulfur Adlayers on the low-index faces of silver. The Journal of Physical Chemistry. 1996;**100**(23):9854-9859

[21] Aloisi GD, Cavallini M, Innocenti M, Foresti ML, Pezzatini G, Guidelli R. In situ STM and electrochemical investigation of sulfur oxidative underpotential deposition on Ag(111). The Journal of Physical Chemistry. B. 1997;**101**(24):4774-4780

[22] Lastraioli E, Loglio F, Cavallini M, Simeone F, Innocenti M, Carlà F, Foresti ML. In situ scanning tunneling microscopy investigation of sulfur oxidative underpotential deposition on Ag(100) and Ag(110). Langmuir. 2010;**26**(22):17679-17685

[23] Perera AGU, Jayaweera PVV, Ariyawansa G, Matsik SG, Tennakone K, Buchanan M, Liu HC, Su XH, Bhattacharya P. Room temperature nano-and microstructure photon detectors. Microelectronics Journal. 2009;**40**(3):507-511

[24] Patil SV, Deshmukh PR, Lokhande CD. Fabrication and liquefied petroleum gas (LPG) sensing performance of p-polyaniline/n-PbS heterojunction at room temperature. Sensors and Actuators B: Chemical. 2011;**156**(1):450-455

[25] Brennan TP, Trejo O, Roelofs KE, Xu J, Prinz FB, Bent SF. Efficiency enhancement of solid-state PbS quantum dot-sensitized solar cells with Al_2O_3 barrier layer. Journal of Materials Chemistry A. 2013;**1**(26):7566-7571

[26] Lachenwitzer A, Morin S, Magnussen OM, Behm RJ. In situ STM study of electrodeposition and anodic dissolution of Ni on Ag(111). Physical Chemistry Chemical Physics. 2001;**3**(16):3351-3363

[27] Medway SL, Lucas CA, Kowal A, Nichols RJ, Johnson D. In situ studies of the oxidation of nickel electrodes in alkaline solution. Journal of Electroanalytical Chemistry. 2006;**587**(1): 172-181

[28] Innocenti M, Cattarin S, Cavallini M, Loglio F, Foresti ML. Characterisation of thin films of CdS deposited on Ag(111) by ECALE. A morphological and photoelectrochemical investigation. Journal of Electroanalytical Chemistry. 2002;**532**(1):219-225

[29] Cecconi T, Atrei A, Bardi U, Forni F, Innocenti M, Loglio F, Foresti ML, Rovida G. X-ray photoelectron diffraction (XPD) study of the atomic structure of the ultrathin CdS phase deposited on Ag(111) by electrochemical atomic layer epitaxy (ECALE). Journal of Electron Spectroscopy and Related Phenomena. 2001;**114-116**:563-568

[30] Blachnik R, Muller A. The formation of Cu_2S from the elements. II. Copper used in form of foils. Thermochimica Acta. 2001;**366**(1):47-59

[31] Beck Tan NC, Balogh L, Trevino SF, Tomalia DA, Lin JS. A small angle scattering study of dendrimer–copper sulfide nanocomposites. Polymer. 1999;**40**(10):2537-2545

[32] Bencista I, Di Benedetto F, Innocenti M, De Luca A, Fornaciai G, Lavacchi A, Montegrossi G, Oberhauser W, Pardi LA, Romanelli M, Vizza F, Foresti ML. Phase composition of Cu_xS thin films: Spectroscopic evidence of covellite formation. European Journal of Mineralogy. 2012;**24**:879-884

[33] Di Benedetto F, Bencistà I, Caporali S, Cinotti S, De Luca A, Lavacchi A, Vizza F, Muniz Miranda M, Foresti ML, Innocenti M. Electrodeposition of ternary $Cu_xSn_yS_z$ thin films for photovoltaic applications. Progress in Photovoltaics: Research and Applications. 2014;**22**: 97-106

[34] Di Benedetto F, Borrini D, Caneschi A, Fornaciai G, Innocenti M, Lavacchi A, Massa CA, Montegrossi G, Oberhauser W, Pardi LA, Romanelli M. Magnetic properties and cation ordering of nanopowders of the synthetic analogue of kuramite, Cu_3SnS_4. Physics and Chemistry of Minerals. 2011;**38**(6):483-490

[35] Innocenti M, Cattarin S, Loglio F, Cecconi T, Seravalli G, Foresti ML. Ternary cadmium and zinc sulfides: Composition, morphology and photoelectrochemistry. Electrochimica Acta. 2004;**49**:1327-1337

[36] Loglio F, Innocenti M, Pezzatini G, Foresti ML. Ternary cadmium and zinc sulfides and selenides: Electrodeposition by ECALE and electrochemical characterization. Journal of Electroanalytical Chemistry. 2004;**562**:117-125

[37] Caporali S, Tolstogouzov A, Teodoro OMND, Innocenti M, Di Benedetto F, Cinotti S, Picca RA, Sportelli MC, Cioffi N. Sn-deficiency in the electrodeposited ternary $Cu_xSn_yS_z$ thin films by ECALE. Solar Energy Materials & Solar Cells. 2015;**138**:9-16

[38] Di Benedetto F, Cinotti S, D'Acapito F, Vizza F, Foresti ML, Guerri A, Lavacchi A, Montegrossi G, Romanelli M, Cioffi N, Innocenti M. Electrodeposited semiconductors at room temperature: An X-ray absorption spectroscopy study of cu-, Zn-, S-bearing thin films. Electrochimica Acta. 2015;**179**:495-503

[39] Wei C, Myung N, Rajeshwar K. A combined voltammetry and electrochemical quartz crystal microgravimetry study of the reduction of aqueous Se(IV) at gold. Journal of Electroanalytical Chemistry. 1994;**375**:109-115

[40] Pezzatini G, Loglio F, Innocenti M, Foresti ML. Selenium(IV) electrochemistry on silver: A combined electrochemical quartz-crystal microbalance and cyclic voltammetric investigation. Collection of Czechoslovak Chemical Communications. 2003;**68**(9):1579-1595

[41] Cavallini M, Aloisi G, Guidelli R. An in situ STM study of selenium electrodeposition on Ag(111). Langmuir. 1999;**15**(8):2993-2995

[42] Russak MA, Reichman J, Witzke H, Deb SK, Chen SN. Thin film CdSe photoanodes for electrochemical photovoltaic cells. Journal of the Electrochemical Society. 1980;**127**(3): 725-733

[43] Liu C-HJ, Wang JH. Spray-pyrolyzed thin film CdSe photoelectrochemical cells. Journal of the Electrochemical Society. 1982;**129**(4):719-722

[44] Kim S, Fisher B, Eisler H-J, Bawendi M. Type-II quantum dots: CdTe/CdSe(core/shell) and CdSe/ZnTe(core/shell) heterostructures. Journal of the American Chemical Society. 2003; **125**(38):11466-11467

[45] Dabbousi BO, Rodriguez-Viejo J, Mikulec FV, Heine JR, Mattoussi H, Ober R, Jensen KF, Bawendi MG. (CdSe)ZnS core−shell quantum dots: Synthesis and characterization of a size series of highly luminescent Nanocrystallites. The Journal of Physical Chemistry B. 1997;**101**(46):9463-9475

[46] Gur I, Fromer NA, Geier ML, Alivisatos AP. Air-stable all-inorganic nanocrystal solar cells processed from solution. Science. 2005;**310**(5747):462-465

[47] Huang Y, Duan X, Lieber CM. Nanowires for integrated multicolor nanophotonics. Small. 2005;**1**(1):142-147

[48] Loglio F, Innocenti M, D'acapito F, Felici R, Pezzatini G, Salvietti E, Foresti ML. Cadmium selenide electrodeposited by ECALE: Electrochemical characterization and preliminary results by EXAFS. Journal of Electroanalytical Chemistry. 2005;**575**(1):161-167

[49] Pezzatini G, Caporali S, Innocenti M, Foresti ML. Formation of ZnSe on Ag(111) by electrochemical atomic layer epitaxy. Journal of Electroanalytical Chemistry. 1999;**475**(2): 164-170

[50] Loglio F, Telford AM, Salvietti E, Innocenti M, Pezzatini G, Cammelli S, D'Acapito F, Felici R, Pozzi A, Foresti ML. Ternary $Cd_xZn_{1-x}Se$ deposited on Ag (111) by ECALE: Electrochemical and EXAFS characterization. Electrochimica Acta. 2008;**53**:6978-6987

[51] Mori E, Baker CK, Reynolds JR, Rajeshwar K. Aqueous electrochemistry of tellurium at glassy carbon and gold: A combined voltammetry-oscillating quartz crystal microgravimetry study. Journal of Electroanalytical Chemistry. 1988;**253**:441-451

[52] Traore M, Modolo R, Vittori O. Electrochemical behaviour of tellurium and silver telluride at rotating glassy carbon electrode. Electrochimica Acta. 1988;**33**:991-996

[53] Forni F, Innocenti M, Pezzatini G, Foresti ML. Electrochemical aspects of CdTe growth on the face (111) of silver by ECALE. Electrochimica Acta. 2000;**45**:3225-3231

[54] Liang YQ, Zhai L, Zhao XS, Xu DS. Band-gap engineering of semiconductor nanowires through composition modulation. The Journal of Physical Chemistry B. 2005;**109**(15): 7120-7123

[55] Takahashi T, Nichols P, Takei K, Ford AC, Jamshidi A, Wu MC, Ning CZ, Javey A. Contact printing of compositionally graded CdS_xSe_{1-x} nanowire parallel arrays for tunable photodetectors. Nanotechnology. 2012;**23**(4):045201

[56] Foresti ML, Milani S, Loglio F, Innocenti M, Pezzatini G, Cattarin S. Ternary $CdS_xSe_{(1-x)}$ deposited on Ag(111) by ECALE: Synthesis and characterization. Langmuir. 2005;**21**(15): 6900-6907

Chapter 6

Pulsed Electrochemical Deposition of $CuInSe_2$ and $Cu(In,Ga)Se_2$ Semiconductor Thin Films

Sreekanth Mandati, Bulusu V. Sarada,
Suhash R. Dey and Shrikant V. Joshi

Additional information is available at the end of the chapter

http://dx.doi.org/10.5772/intechopen.71857

Abstract

$CuInSe_2$ (CIS) and $Cu(In,Ga)Se_2$ (CIGS) semiconductors are the most studied absorber materials for thin films solar cells due to their direct bandgap and large absorption coefficient. The highly efficient CIGS devices are often fabricated using expensive vacuum based technologies; however, recently electrodeposition has been demonstrated to produce CIGS devices with high efficiencies and it is easily amenable for large area films of high quality with effective material use and high deposition rate. In this context, this chapter discusses the recent developments in CIS and CIGS technologies using electrodeposition. In addition, the fundamental features of electrodeposition such as direct current, pulse and pulse-reverse plating and their application in the fabrication of CIS and CIGS films are discussed. In conclusion, the chapter summarizes the utilization of pulse electrodeposition for fabrication of CIS and CIGS films while making a recommendation for exploring the group's unique pulse electroplating method.

Keywords: pulse electrodeposition, semiconductors, thin films, copper indium gallium selenide, photoelectrochemical cells, solar cells

1. Introduction

The present day's global energy production is mostly accomplished from the fossil fuels; however, the inherent problems associated with the use of fossil fuels such as their limited availability and the environmental issues force the mankind to look for alternative solutions for future energy supply. The need to develop renewable energy resources has come to the forefront of discussion. Photovoltaics (PV) is an emerging field and one of the choices for major means of future energy-harvesting. The efficiencies of PV conversion depend on the properties of absorber layer, which is the heart of the solar cell. Silicon has been the foundation of the photovoltaics industry due to its well-known properties, high abundance and well-studied technological aspects of the growth. However, it requires expensive manufacturing technologies such as growing and sawing of ingots. Hence, current trend in photovoltaics requires the development of high performance inexpensive solar absorber materials that can serve in the long term as viable alternatives to the single crystal silicon technology. Among various technologies, $CuInSe_2$ based solar cells are the most studied and leading candidates to realize commercialization with efficiencies close to 23% [1].

2. Copper indium selenide (CIS) and copper indium gallium selenide (CIGS)

CIS is a ternary semiconductor belonging to the I-III-VI class, crystallizes to a chalcopyrite structure, possesses a direct bandgap of 1.04 eV and an absorption coefficient of $\approx 10^5$ cm^{-1} [2, 3]. The CIS-based solar cells exhibit excellent chemical stability, stability with time and doping versatility. Absorber layer is the key element of solar cells, which is produced mainly from the p-type semiconductor in thin films solar cells. The electrical properties of Cu ternary semiconductors are determined by native defects [4]. There are three possible electrically actives defects namely, vacancies, interstitials and antisite defects [5–7]. It is these defects which determine the nature of the conductivity of CIS films whether n type or p type. Intrinsic copper vacancies (V_{Cu}) and copper on indium antisite defects (Cu_{In}) are the electrically active defects for a typical p-type CIS film. On the other hand, intrinsic selenium vacancies (V_{Se}) and indium on copper antisite defects (In_{Cu}) makes the CIS n-type. The material with Cu rich composition is not preferred mainly because of the formation of copper selenide (Cu_2Se). Cu_2Se being highly conductive, shorts out the junction. Adding more Indium than copper reduces the formation of Cu_2Se but it causes other defects like V_{Cu} and In_{Cu} which are compensating in nature [8]. Hence, the copper to indium ratio (Cu/In) is always maintained around unity. Samples with p-type conductivity are grown if the material is Cu-poor and is annealed under high Se vapor pressure, whereas Cu-rich material with Se deficiency tends to be n-type. CIS films when suitably manufactured tend to be p-type because of the low energy of formation of copper vacancies which give the material its conductivity [4, 7, 9]. CIS solar cells yielded relatively lower open circuit potentials due to its small bandgap. This limitation is overcome by adding controlled amounts of gallium to replace indium in the CIS structure [10]. The band gap of $CuIn_{1-x}Ga_xSe_2$ varies according to the equation [11].

$$E_g = 1.011 + 0.664x + 0.249x(1 - x) \quad (1)$$

Depending on the [Ga]/[In + Ga] ratio, the bandgap of CIGS can be varied continuously between 1.02 and 1.68 eV [10, 12]. The addition of about 30% Ga in CIS increases the bandgap to 1.2 eV which has a closer match with the AM 1.5 solar spectrum [12, 13]. Addition of Ga not only increases the band gap but also has other beneficial effects such as improving the adhesion of the film to the Mo substrate, increased carrier concentration, etc. Though it is beneficial to add Ga to improve the properties of CIS, there is a limit to which it serves favorable.

Higher Ga content of 40% has a detrimental effect on the device performance, because it negatively impacts the transport properties of the CIGS absorber film. The current, high-efficiency devices are prepared with bandgaps in the range 1.20–1.25 eV, which corresponds to a Ga/(In + Ga) ratio between 25 and 30% [14–18].

The fabrication methods often are the big difference in industrializing a technology based on a materials system. To determine the most promising technique for the commercial manufacturing of modules, the foremost criteria are: (i) low cost, (ii) scalability, (iii) reproducibility and (iv) manufacturability. Several vacuum techniques including co-evaporation, sputtering, molecular beam epitaxy, pulsed laser deposition, etc., have been investigated for the formation of CIGS thin-films. The most successful technique for deposition of CIGS absorber layers for highest efficiency small area cells is the co-evaporation of elements from multiple sources where Se is offered in excess during the process [19]. Although thin-film CIGS solar cells with power conversion efficiencies over 23% have been demonstrated, the vacuum-based processes used therein pose cost and technological barriers in the production of PV modules. With this requirement in mind, development of low-cost methods to fabricate CIGS has become an intensively pursued goal and a variety of solution-based approaches have been demonstrated. Non-vacuum approaches, include electro or electroless-deposition, chemical bath deposition, particulate processes and coating involving molecular precursors are being explored. The efficiency gap between vacuum and non-vacuum deposited CIGS has been reduced in past years and processes from the above categories have now reported cells with efficiencies of 15–17%, thereby, showing promise for commercialization. Among these methods, electrochemical deposition is the most extensively explored technique for the deposition of CIGS absorber layers and has witnessed high efficiency devices [20–23], and will be discussed in detail in the following section.

3. Electrodeposition

The process of electro-reduction of precursor ions onto an electrode, substrate of interest, from an appropriate electrolyte by the utilization of electric current or potential between two electrodes is termed as electrodeposition. The properties of the deposit depends on several parameters including ionic concentrations, the electrodes used, the pH of the electrolyte, the temperature, the type of substrate, the stirring rate, the deposition voltage and the time of deposition. Electrodeposition is a promising approach to fabricate the absorber layers in thin films solar cells due to following advantages:

- It is a low cost solution based approach and avoids expensive vacuum technology
- High deposition rate on larger areas with lateral uniformity
- Use of low cost starting materials, e.g., low purity salts and solvents can be purified with application of small voltage before the electrodeposition
- Uses low temperature for deposition
- Deposition of films on a variety of shapes and forms of wires, tapes, coils, and cylinders

- Effective material use as high as 98% with minimum waste generation
- Control over the composition of individual elements by varying deposition parameters
- Extrinsic doping of semiconductors with appropriate elements by utilizing small concentrations of dopant precursor in electrodeposition bath
- Bandgap engineering either by varying the contents of individual elements in step-by-step approach or by using combinatorial approach

Owing to these advantages, electrodeposition satisfies all the necessary criteria for the research and development of PV solar cells. In addition, electrodeposition involves two fundamental approaches such as direct current and pulse plating wherein pulse plating further comprises pulse and pulse-reverse features. These features are unique in themselves that they provide additional process control variables and make electrodeposition an attractive tool for the deposition of semiconductors [24].

3.1. Direct current (DC) Electrodeposition

Direct current electrodeposition is the conventional method wherein a constant current or potential is applied continuously during the deposition to coat the desired materials. **Figure 1** shows the schematic of the direct current applied and the typical growth process corresponding to it. DC technique effectively has two variables, namely, applied potential/current and time of deposition while the precursor concentration and electrolyte pH are the common variables. The morphology, composition and thickness of the deposit can be altered by varying these parameters. In DC, the continuous use of constant potential/current leads to deposition of films without any relaxation leading to the growth of existing nuclei instead of generating new nucleation sites thereby resulting in a rough and porous deposit. In addition, hydrogen evolution reaction (HER) competes along with the deposition of desired materials as often aqueous electrolytes are used. The HER not only influences the current efficiency of the technique but also the deposited film properties. Use of additives in conventional DC deposition has improved the morphology of the deposited films. Despite having several disadvantages, DC electrodeposition is still a leading technique for the production of single element deposits and binary alloys.

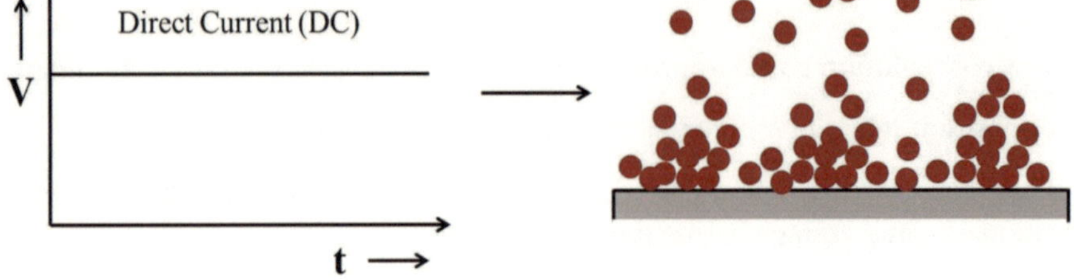

Figure 1. Schematic of direct current electrodeposition and expected growth process of the deposit.

http://dx.doi.org/10.5772/intechopen.71857

3.2. Pulse electrodeposition

In pulse electrodeposition (PED), current/potential is applied in the form of modulated waves as shown in **Figure 2**. Compared to DC, PED offers additional process control variables like pulse on-time (t_{on}) and off-time (t_{off}). Often, the variation in t_{on} and t_{off} is expressed using a common parameter known as duty cycle, which is defined by the equation:

$$Duty\ Cycle\ (\%) = \frac{t_{on}}{t_{on} + t_{off}} \times 100 \tag{2}$$

The precise variation in duty cycle provides the control over electrochemical processes by affecting the diffusion layer, grain size and nucleation. Usually in electroplating a negatively charged layer is formed around the cathode which gets charged to a known thickness and prevents the ions from the bulk. In DC, this charged double layer prevents the passage of ions toward the cathode thereby affecting the features of the deposit and also causes the inhomogeneity in the concentration of ions in the electrolyte. However, in PED, the charged diffusion layer gets discharged and helps easier passage of the ions onto the cathode as the output is periodically turned off. Migration of ions to depleted areas in the bath during off-time makes the even distribution of ions for their easy availability during on-time. The presence of off time aids in the relaxation and the rearrangement of deposited atoms leading to the possibility of new nucleation sites during the subsequent deposition. This, in turn, not only improves the uniformity of deposition but also reduces the porosity and roughness of the deposit (see **Figure 2**). In addition, the entrapped hydrogen and impurities during the deposition diffuse out during the relaxation time. The additional process variables in PED ease the process of optimization for complex stoichiometric ternary/quaternary systems as the control over individual elemental composition is far better with the variation in duty cycle compared to DC technique.

3.3. Pulse reverse electrodeposition

Pulse reverse electrodeposition contains short anodic pulses alongside the cathodic as seen in PED (See **Figure 3**). This small anodic pulse is advantageous since it contributes to the electro-oxidation of the top layer from the deposited film thereby aids in smoothening of the deposit, removing the impurities as well as the entrapped hydrogen. The anodic pulse plays a major role in the deposition of systems like CIS and CIGS, wherein an undesired

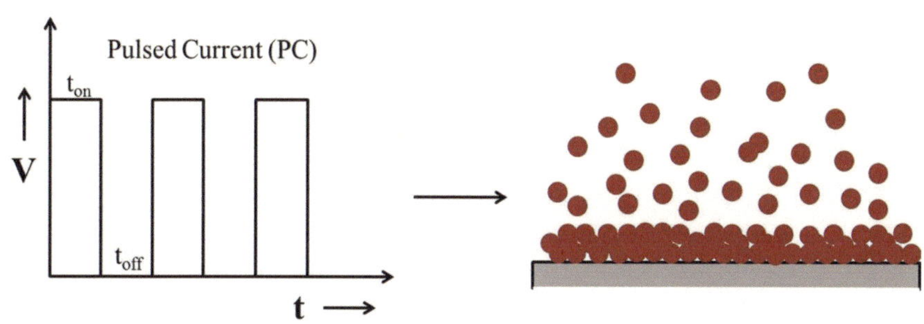

Figure 2. Schematic of pulse electrodeposition and expected growth process of the deposit.

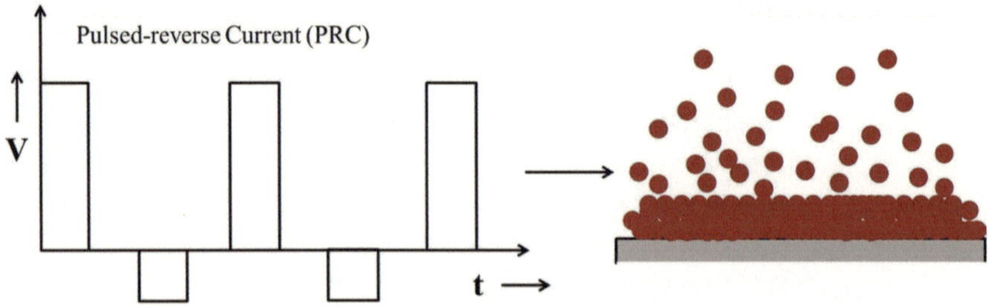

Figure 3. Schematic of pulse reverse electrodeposition and expected growth process of the deposit.

secondary Cu-Se phase exists on the surface. With the appropriate control of the anodic pulse, excess copper and the undesired phases can be easily eliminated from the deposited films thereby, forming a phase-pure CIS/CIGS. Similar to PED deposition, the variation in pulse parameters is expressed using a common parameter known as duty cycle, which is defined by the equation:

$$Duty\ Cycle\ (\%) = \frac{t_c}{t_a + t_c} \times 100 \qquad (3)$$

3.4. Electrodeposition of ternary/quaternary chalcopyrites

The goal of the electrodeposition (ED) is to assure an adherent, compact and a laterally uniform film with the desired stoichiometry. Lateral compositional uniformity is essential over large areas for commercialization of devices. The properties of the deposit solely depend on the control of individual parameters during ED such as electrode material, precursor concentration in electrolyte, applied potential/current and temperature. For single metallic systems like Cu or Zn these properties are well understood. However, deposition of CIGS contains multiple elements including a chalcogen making the ED process substantially complex. Despite that electrochemical deposition appears to be a promising technique for the low-cost solution preparation of CIGS semiconductors [25–27]. Indeed, electrochemical deposition has been widely investigated for CIGS deposition since the pioneering work by Bhattacharya et al. in 1983 [28]. Since then, more than 350 publications have been devoted to the electrochemical preparation of CIGS and several review papers have appeared [25, 29, 30], making electrodeposition the most intensely studied non-vacuum deposition method for CIGS. In light of this, this section reviews and summarizes the previously published work.

The first electrochemical approach to deposit polycrystalline CIS was reported by Bhattacharya in 1983, wherein Cu, In and Se were simultaneously deposited from an acidic solution [28]. Quickly after the first report, several approaches have been proposed for the synthesis of CIS thin films [31–33]. It is of general interest to perform a simultaneous codeposition of all three elements in achieving stoichiometric chalcopyrite CIS films. Such an approach was first initiated by Bhattacharya [28], which is by far the most investigated case as it involves only one electrochemical process, often termed as one-step electrodeposition. Typical electrochemical reactions involving the reduction of elements and formation of CIS/CIGS phase are detailed in [34].

One-step electrodeposition of CIS is usually carried out in an aqueous solution often containing chloride/sulfate precursors of Cu^{2+} or Cu^{+}, In^{3+}, and SeO_2/H_2SeO_3. The deposition solution often contains a complexing agent in order to shift the reduction potentials of Cu and In closer together to improve the film quality. Complexing agents such as citric acid/citrate [35], ammonia [28], triethanolamine [28, 36], thiocyanate [37], etc., are used during the one-step electrodeposition of CIS thin-films. In addition, a supporting electrolyte such as chloride (LiCl [22, 34, 38, 39] or NaCl [40]) or sulfate (K_2SO_4 [41–43]) is added which results in an improved conductivity of the electrolyte leading to easier mobility of the precursor ions. Often amorphous or poorly crystalline CIS films were observed from electrodeposition which contains frequently degenerate $Cu_{2-x}Se$ phases that are detrimental to the device performance [30, 37]. Also, Cu-rich films have generally larger grain sizes than stoichiometric or In-rich films. Due to these reasons, the electrodeposited CIS films often required to be annealed in a selenium atmosphere to correct for the stoichiometry and improve crystallinity.

Incorporation of Ga into the CIS thin-films, to improve the desired properties, was a challenging task for long time for the formation of quaternary CIGS thin-films. However, this bottleneck has been overcome in the recent past. Several researchers have reported the successful incorporation of Ga in the films up to desired range of amounts (6–8 at. %) for the preparation of high efficiency cells [22, 34]. Bhattacharya et al. were the first to report the insertion of Ga from a chloride bath, but to a very low content $Ga/In \approx 0.1$, wherein a superimposed alternating voltages has been used at 20 kHz [44]. But the breakthrough for the incorporation of Ga has been realized a little later when the group had used a pH buffer in the chloride bath, also known as Hydrion buffer ($pH = 3$) consisting of sulfamic acid and potassium hydrogen phthalate. The deposition potential was kept constant while the solution composition has been varied to realize the real possibility of incorporating Ga with the ratio Ga/In from 0.3 to 0.7. This process demonstrated the formation of CIGS films over a wide range of compositions suitable for efficient solar cells by one-step electrodeposition technique. CIGS layers generally deposited from the above mentioned electrodeposition technique often used an additional PVD step to achieve the required composition to form stoichiometric films [45, 46]. Bhattacharya et al. demonstrated a cell efficiency of 9.4% by using a similar PVD step to improve the composition of In and Ga in the as-deposited CIGS thin-films [39]. Valderrama et al. explored a similar electrodeposition technique followed by PVD step to achieve stoichiometric chalcopyrite CIGS films and additionally demonstrated the use of CIGS films to produce hydrogen by the use of photoelectrochemical testing of the films in H_2SO_4 [46]. Avoiding the formation of secondary phases was essentially a key point to achieve high quality CIGS thin-films and hence, higher efficiency. The formation of secondary phases was successfully prevented by the pretreatment of Mo substrate wherein a 1 min pre-deposition of CIGS was performed and a multi-potential deposition regime was employed to obtain crack-free CIGS layers [47].

In the similar context, addition of Ga seems to cause morphology related problems, often a concern during the electrodeposition of CIGS films. Fernandez et al. varied the concentration of precursor solution systematically and achieved a better control over composition and morphology of the CIGS films [39]. The presence of cracks in Ga-containing layers is often a serious problem [39, 47], though it can be reduced through the use of alcohol–aqueous solutions [48] or supporting electrolytes such as LiCl or Li_2SO_4 with gelatin as brightening additive [49]. Complexing agents such as citric acid/citrate [50, 51], thiocyanate [52], sodium sulfamate [53], sulfosalicylic acid [54], etc., were often used to improve the composition and morphology of the CIGS films. These additives form complexes with the metal ions in the solution such as Cu, thereby resulting in controlled deposition rates and hence the morphology [55, 56]. Good quality CIGS thin films were also prepared from sulfate–citrate and chloride–citrate solutions [53, 57, 58] and control of the optical band gap by increasing the Ga content in the films was demonstrated [59].

To overcome the difficulty of In and Ga incorporation, alternative strategies to co-electrodeposition have been developed, often involving the deposition of stacked elemental layers or else deposition of alloys, followed by a selenization or sulphurization treatment to provide all of the chalcogen [21, 34]. Deposition of Cu-Ga [60] and Cu-In-Ga [52] alloys has been demonstrated, with the latter approach leading to 4% efficiency cells. Solopower explored the layer by layer electrodeposition to form stoichiometric CIGS thin-films [21]. The process also used several varieties of complexing agents, organic additives, etc., to correct the composition and improve the morphology of the films. In addition, current densities were varied during deposition and annealing temperatures were optimized to achieve high quality compact large grained stoichiometric chalcopyrite CIGS films. Devices and modules were fabricated using these films which had shown an efficiency of 15.36% on an area of 5.34 cm^2 and 13.4% on an area of 3.8 m^2. Most recently, electrodeposited CIGSSe devices have yielded conversion efficiencies upto 17.3% wherein electrodeposited Cu-In-Ga stack is rapid thermal treated in elemental selenium and sulfur atmosphere. Some of the recent notable reports with direct current electrodeposited CIGS solar cells are summarized in **Table 1**.

In addition to the synthesis of CIS and CIGS thin films by DC electrodeposition, PED technique has also been studied. Owing to the advantages of PED discussed in Section 3.2, it results in superior quality thin-films and enables one to avoid additional steps of deposition, post deposition treatments and the use of complexing agents. Kang et al. reported the preparation of CIS thin-films by pulse-reverse electrodeposition followed by selenization wherein stoichiometric CIS films with rough surface morphology exhibited an efficiency of 1.42% [61]. Li et al. employed a square wave modulated by a bell-like wave during the pulse plating for the fabrication of CIS thin-films. The study reported well adherent chalcopyrite CIS films with a uniform morphology [35]. A three-step pulse electrodeposition method was used for the fabrication of CIS thin-films reporting a mixture of phases such as Cu-Se, In-Se, CIS, etc., as confirmed from Raman and optical studies of the samples [62]. Valdes et al. employed different potentials during the pulses which resulted in chalcopyrite p-type CIS thin-films with different morphologies and compositions [63]. Murali et al. prepared CIS thin-films using pulse electrodeposition with varied duty cycle from 6 to 50% and reported the p-type phase pure CIS films with resistivities in the range of 1–10 ohm-cm [64]. Hu et al. employed the deposition of CIS films by pulse electrodeposition wherein multi potentials were used to control the composition of the films which resulted in improved deposition uniformity without any secondary phases [65]. Similar reports on the compositional control of CIS thin-films by the variation of pulse parameters have appeared in the recent past [66–69].

Fu et al., have explored different plating techniques including DC, pulse and pulse reverse electrodeposition for the fabrication of CIGS thin films and reported the elimination of undesired secondary phases like $Cu_{2-x}Se$ to obtain single phase pure chalcopyrite CIGS thin films [70]. Liu et al., have employed the PED with the variation of duty cycle to remove the excess In and to avoid In-Se compounds during the deposition for the preparation of single phase CIGS

Absorber material	Substrate	Preparation method	Reported efficiency/ photocurrent (cell area in cm^2)	Remarks	Reference
CIGSe	Mo/glass	Cu-In-Ga oxide precursor followed by thermochemical reduction and selenization	12.4% (0.1)	A new approach by one-step electrochemical deposition of metal oxides followed by thermochemical reduction and selenization	[82]
CIGSe	Mo/glass	Stacked Cu/In/ Ga followed by selenization	11.7 (0.4)	Step-by-step electrodeposition of Cu/In/Ga followed by selenization	[22]
CIGSe	Mo/glass	Stacked Cu/In/Ga layers followed by selenization	12.6 (0.1)	Impact of thickness is studied. 8.7% efficiency is achieved for an ultrathin 370 nm film of CIGS	[83]
CIGSe and CIGSSe	Mo/stainless steel foil	Four step electrodeposition of CIG layers followed by selenization	15.36% for CIGSe (5.4) 13.4% for CIGSSe (3824.6, submodule)	Development of roll-to-roll electrodeposited CIGS solar cells on flexible stainless steel substrates	[21]
CIGSe and CIGSSe	Mo/glass	Electrodeposited CuInGa layers annealed using Rapid thermal Processing under S and Se vapors	14.1% for CIGSe (0.5) 15.8% for CIGSSe (0.5)	Rapid thermal processing of electrodeposited CIG layers under S and Se atmosphere	[84]
CIGSSe	Mo/glass	Electrodeposited precursors followed by rapid thermal processing in S and Se atmosphere	10.6 (0.1) for CISSe 9.9 (0.1) for CISe	Sequentially electrodeposited copper indium precursors are annealed by RTP under S and Se atmosphere	[26]
CIGSSe	Mo/glass	Electrodeposited stacked Cu/In/Ga layers are annealed by RTP under S and Se atmosphere	13.8 (0.09)	All solution processed CIGSSe solar cells are made wherein i-ZnO/AZO window layer is deposited using solution based approach.	[85]
CIGSe	Mo/glass	Electrodeposited CIGSe layers followed by physical vapor deposition of In, Ga, Se and selenized	15.4% for ED CIGSe (0.4) 12.4% for EL CIGSe (0.4)	Physical vapor deposition of In, Ga and Se was performed on electrodeposited and electroless deposited CIGS layers for compositional adjustment	[23]

Absorber material	Substrate	Preparation method	Reported efficiency/ photocurrent (cell area in cm^2)	Remarks	Reference
CIGSe	Mo/glass	Three stage electrodeposition (CIGSe/Cu/In) process followed by annealing under Se atmosphere	10.9% (0.4)	Electrodeposited precursor layers are prepared in a three stage sequential route wherein Cu and In second and third layers are deposited on CIGSe layer	[38]
CIGSe CISe	Mo/glass	Electrodeposited CIGSe is annealed under selenium atmosphere and spray coated CIS nanoparticles followed by selenization	Photocurrent of 1 mA/cm^2 for CISe and for CIGSe in ethyl viologen perchlorate electrolyte	CISe films were deposited from spray coating of nanoparticles and selenized while CIGSe were electrodeposited and selenized	[86]

Table 1. Notable reports on direct current electrodeposited CIS/CIGS with reported efficiencies.

thin films [58]. In one of the most successful efforts of fabrication of CIGS by pulse electrodeposition, Bi et al. have effectively utilized the parameters in electrodeposition to demonstrate CIGS solar cells with conversion efficiencies upto 10.39% [71] and 11.04% [72]. Cu-In-Ga metal precursors were electrodeposited by pulse current method wherein the charge density was chosen such that the desirable thickness of each layer is achieved. Electrodeposited metal stack was annealed in a three step process to achieve dense large grained CIGS layer which yielded high conversion efficiencies. **Table 2** summarizes the recent literature reported for CIGS thin films fabrication using pulse electrodeposition.

Materials	Substrate	Preparation method	Notable inference	Remarks	Reference
CIGSe	Mo/glass	Stacked Cu/In/Ga by DC and PC methods followed by selenization	11.04% efficiency for pulse plated CIGSe (0.34) 8.18% efficiency for direct current deposited CIGSe (0.34)	Direct current and pulsed current plating are explored for step by step Cu/In/Ga layers and pulse plated devices exhibited higher efficiency	[72]
CIGSe	Mo foil	DC, Pulse and pulse reverse plating of CIGS films followed by annealing in Ar	CIGS exhibited p-type and dark I-V curves are recorded	Use of pulse reverse potential eliminated the undesired CuSe phase	[70]
CISe	Mo/glass	Pulse reverse electrodeposition of CISe followed by selenization	Efficiency of 1.42% on 0.4 cm^2 cell	Pulse reverse plating is employed to make CISe with device quality features	[61]
CIGSe	Mo foil	Pulse electrodeposition of CIGSe followed by annealing	PEC curves are recorded indicating photoactivity of CIGSe films	Variation in duty cycle is adopted and growth mechanism is studied	[58]

Materials	Substrate	Preparation method	Notable inference	Remarks	Reference
CIGSe	Mo/glass	DC and pulse reverse plating of CIGS followed by selenization		Variation in duty cycle in pulse reverse plating yielded device quality CIGS films	[87]
CIGSe	Mo/glass	Step by step pulse electrodeposition of Cu/In/Ga films followed by selenization	10.39% (0.34)	Optimization of pulse conditions for each layer of Cu, In and Ga during pulse electrodeposition	[71]
CIGS	ITO	Pulse electrodeposition of CIGS followed by annealing	Photocurrents in μA range are observed	Duty cycle has been optimized for device quality CIGS films	[88]
CISe	Mo foil	Pulse electrodeposition of CIS followed by annealing	PEC performance reported	A novel nanoflake morphology has been reported while obtaining stoichiometric CISe with optimization in duty cycle	[76]
CIGSe	Mo foil	DC and pulse plating of CIGSe films followed by annealing	PEC performance reported	DC and pulse plated CIGS films are compared. Optimization in pulse plating improves the morphology and crystallinity of CIGSe	[75]
CIGSe	Mo/glass	Two stage pulse electrodeposition of CuGaSe and In films followed by annealing	PEC performance reported	A sequential pulse electrodeposition approach is proposed so as to minimize the utilization of In concentration during electrodeposition of CIGSe films	[73]

Table 2. Summary of recent reports on pulse electrodeposited CIS/CIGS for application in thin films solar cells.

Despite having several advantages, pulse and pulse reverse electrodeposition techniques are underutilized for the fabrication of CIGS thin films. Considering the additional process variables these techniques offer, thin films photovoltaic community might consider exploring these processes for fabricating high quality CIGS films. Our group has explored these features while developing a simplified pulsed electrodeposition technique for the fabrication of CIGS thin films and is discussed below.

4. Pulsed electrodeposition of CIS and CIGS films: case studies

A new simplified pulse electrodeposition approach for the fabrication of CIS and CIGS films with compact and nanostructured morphologies was developed [73–76]. The approach utilizes a two electrode system by avoiding the conventional third reference electrode for the deposition of thin films. The commonly used expensive platinum counter electrode is substituted by high purity graphite electrode thereby cutting down the cost. The technique by virtue of utilization of pulse plating avoids the use of any additives and complexing agents during electrodeposition while obtaining the stoichiometric films in a single step thereby eliminating multi-step deposition. The films are annealed under Ar atmosphere while avoiding the conventional selenium atmosphere making it an environmental friendly approach. In addition, a novel strategy has been developed to address the scarcity of Indium wherein In precursor has been effectively minimized compared to existing reports. Typical pulse electrodeposition set-up, applied pulse voltage wave form and the corresponding current density are depicted in **Figure 4**.

Since the approach avoids the use of reference electrode and employs a two-electrode system, prior to electrodeposition, cyclic voltammetry of the CIGS system using a two and three-electrode system are studied as shown in **Figure 5**. The reduction peak of CIGS is shifted from −0.8 to −1.1 V from three to two-electrode systems. Apparently, the potential shift is approximately same as the standard electrode potential of SCE vs. normal hydrogen electrode (NHE). However, experiments with different systems might be required to verify the same. The inference is utilized to adopt higher deposition potentials in the pulsed electrodeposition of CIS and CIGS films.

4.1. Study I: pulse electrodeposition of $CuInSe_2$ films

The pulse electrodeposition of $CuInSe_2$ (CIS) films is performed from a bath containing metallic chlorides of Cu, In and Ga while selenous acid is used as selenium precursor dissolved in pH 3 Hydrion buffer. The final pH of the bath is maintained around 2.0–2.5. Additives and complexing agents are completely avoided in this study. CIS films were electrodeposited using a deposition potential of −1.5 V while the variation in duty cycle has been studied. The applied pulse-voltage and the corresponding current density curves from **Figure 6** note that there is a small positive current density during the pulse off-time though no voltage is applied. This could be due to the presence of an electric double layer at cathode-electrolyte interface forming a capacitor of molecular dimension [58, 77]. The duty cycle for the application of pulses is varied in the range of 17–67% (varied off-time with fixed on-time). The PED technique employed in

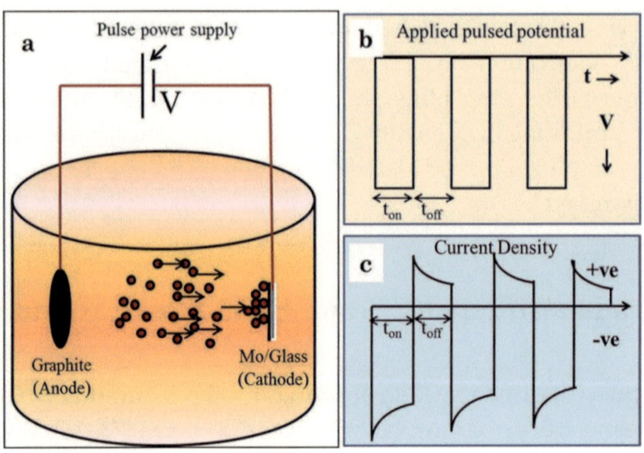

Figure 4. (a) Pulse electrodeposition set-up, (b) applied deposition voltage waveform and (c) corresponding current density employed for pulse electrodeposition of CIS and CIGS thin films.

http://dx.doi.org/10.5772/intechopen.71857

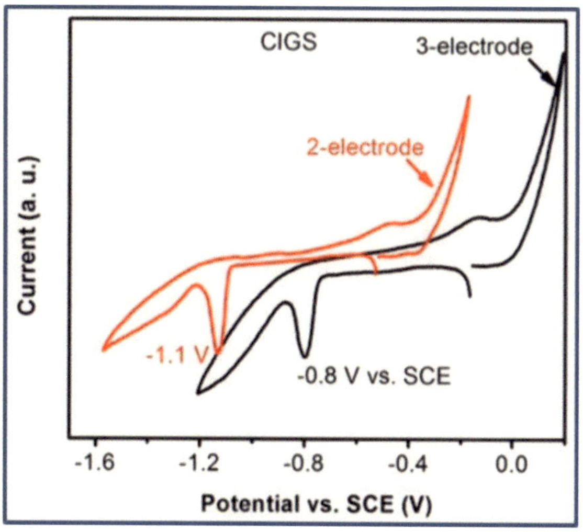

Figure 5. Cyclic voltammogram of Cu-In-Ga-Se system using a three-electrode and two-electrode configuration.

the present study did not result in any disruption and dissolution of the deposited film into the electrolyte, which generally happens with higher deposition voltages as previously reported [78]. The PED deposited CIS films are annealed at 550°C for 30 min under Ar atmosphere.

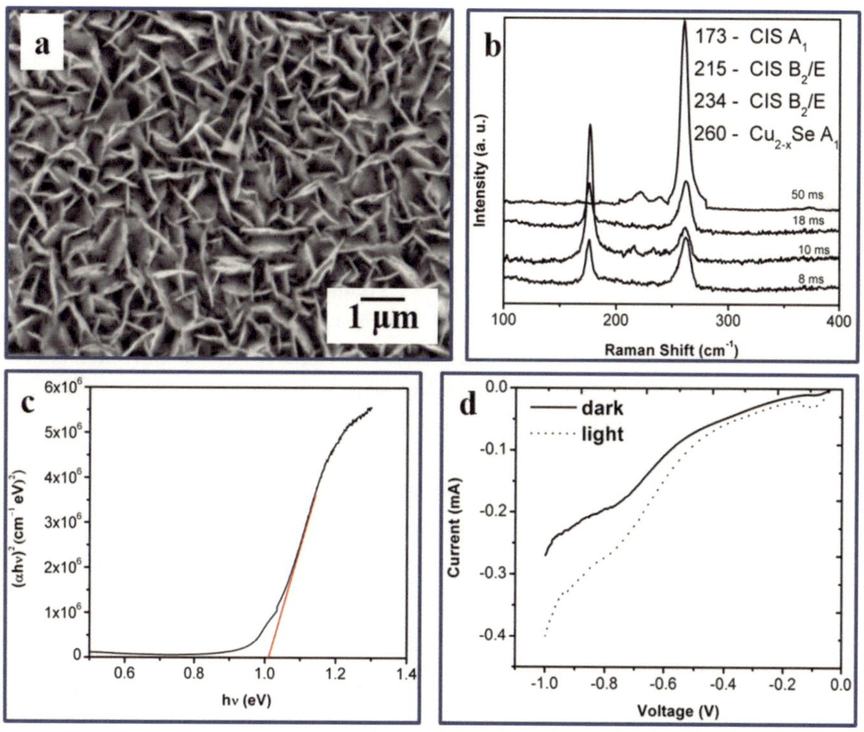

Figure 6. (a) Nanoflake morphology, (b) Raman spectra, (c) Tauc's plot and (d) photoelectrochemical I-V characteristics of pulse electrodeposited and annealed CIS films.

The compositional analysis indicated that with an increase in the pulse off time from about 5 to 50 ms during electrodeposition, the relative content of In in the film decreases from almost 40 to 2 at. %. This is due to the unintended positive current density observed in **Figure 4c**, which oxidizes the elements with least electronegativity from the deposited film leading to the dissolution of corresponding element into the electrolyte. A duty cycle of 50% has been considered optimal for obtaining stoichiometric CIS films. Morphological analysis reveal that the CIS films deposited with optimal conditions possess novel nanoflake-like architectures (see **Figure 6a**). Such morphology is expected to be advantageous since they possess high surface area of the film thereby causing improvement in absorption of light and photoresponse. This is also expected to be advantageous at the device stage since it facilitates increase in the p-n junction interface area of the solar cell which directly influences its performance. Additionally, flake like crystallite structure is favorable to increase current carrier concentration, electron transmission and thus induce the generation of photocurrent [79]. Another interesting aspect from this study is the effective reduction in the secondary phases during the electrodeposition of CIS/CIGS films. The variation in duty cycle explored in the present study yielded the films with reduced copper selenide phase for optimal conditions. Raman spectra of the electrodeposited and annealed CIS films are shown in **Figure 6b**. The spectra contain well-defined peaks at wavenumbers of 172, 215, 234 and 260 cm^{-1} and are attributed to the A_1, B_2 and E modes of CIS films and the peak at 260 cm^{-1} is attributed to the A_1 mode of $Cu_{2-x}Se$. As the spectra reveal the optimized condition with relative Cu/In ratio close to 1, shows least intense peak for copper selenide thereby affirming its reduction. Bandgap of CIS films as obtained from optical absorption spectroscopy is 1.02 eV (from **Figure 6c**), which matches well with the theoretical value of 1.04 eV for stoichiometric chalcopyrite CIS films [80]. Photoelectrochemical performance of CIS films as shown in **Figure 6d** confirms the photoactivity and p-type conductivity.

4.2. Study II: comparison of direct current and pulse electrodeposited CIGS films

CIGS thin films are deposited using direct current (DC) and pulsed current (PC) methods from a bath of metal chlorides and selenous acid in pH 3 Hydrion buffer devoid of additives and complexing agents. **Figure 7a** and **b** shows the surface morphologies of DC and PC plated CIGS films. DC deposited CIGS films exhibited porous and rough morphology with finer spherical particles while dense uniform films with coarser spherical particles are observed for PC deposited CIGS. PC method can produce relatively more homogeneous surface with good adhesion to the substrate because the rate-determining step of the deposition process is controlled by a mass-transfer process. Relaxation during the pulse off-time in the PC electrodeposition not only allows the diffusion of ad atoms but also facilitates the formation of new nucleation sites thereby leading to the homogeneous and compact morphology unlike the DC deposition wherein building-up of material takes place at same nucleation sites leading to the roughness of the film. The stoichiometry of the DC and PC electrodeposited CIGS films is determined, from EDS, to be $Cu_{1.10}In_{0.54}Ga_{0.23}Se_{2.13}$ and $Cu_{0.98}In_{0.73}Ga_{0.25}Se_{2.03}$, respectively, indicating copper rich composition in DC CIGS films.

Figure 7c shows the XRD patterns of annealed DC and PC electroplated CIGS films, which show the preferred orientation corresponding to (112) and other peaks to (211), (220), (312)

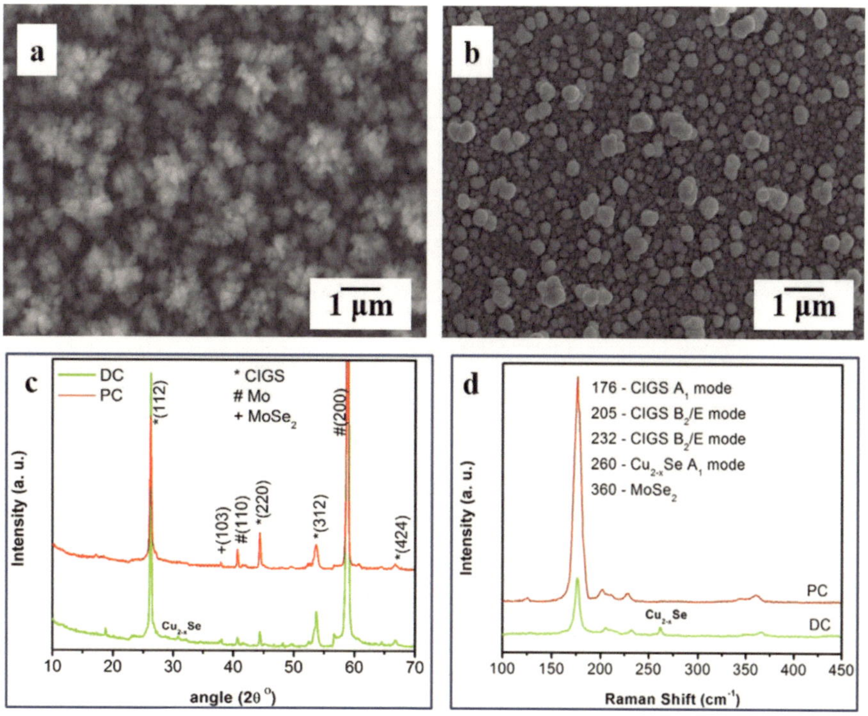

Figure 7. (a) and (b) surface morphologies, (c) XRD patterns and (d) Raman spectra of direct current and pulse electrodeposited CIGS films.

and (424) for CIGS are observed thereby confirming the presence of chalcopyrite CIGS phase. Additionally, Mo substrate and $MoSe_2$ peaks are observed [81]. XRD pattern also reveals the presence of copper selenide in DC electroplated CIGS while it is absent in PC plated films indicating the phase-pure CIGS formation by PC approach. **Figure 7d** shows the Raman spectra of the DC and PC electrodeposited CIGS thin films and contain A_1, B_2 and E modes of the CIGS at 176, 205 and 232 cm^{-1}, respectively. In addition, a less intense peak corresponding to A_1 mode of $Cu_{2-x}Se$ at 260 cm^{-1} is found in case of DC electrodeposited CIGS thin film. The copper rich composition in DC plated CIGS films facilitated the formation of the $Cu_{2-x}Se$ phase, which is generally dispersed on the surface [84, 87]. PC electrodeposition with suitable optimization of parameters aided control over the composition of elements and eliminated $Cu_{2-x}Se$ phase. The bandgap of direct current and pulse plated CIGS films is obtained from Tauc's plots as shown in **Figure 8a**, which are determined to be 1.21 and 1.29 eV, respectively. The photoelectrochemical J-V characteristics of annealed DC and PC electrodeposited CIGS thin films studied in 0.5 M Na_2SO_4 are shown in **Figure 8b**. Increase in cathodic current with potential confirms the p-type conductivity. Also, the PC plated CIGS films show lower dark current and higher photocurrent compared to DC plated films, which could be attributed to the dense morphology with stoichiometric chalcopyrite CIGS without secondary phases. **Figure 8b** shows the amperometric J-t curve confirming the photoactivity of CIGS under chopped illumination.

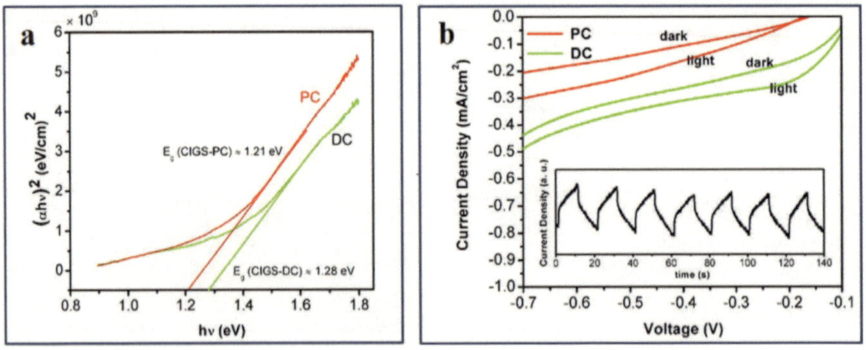

Figure 8. (a) Tauc's plots and (b) photoelectrochemical J-V characteristics of direct current and pulse electrodeposited CIGS films.

4.3. Study III: sequential pulsed electrodeposition of CIGS thin-films

The sequential technique has been proposed to essentially minimize In precursor to address the scarcity of In. The usage of In in the growing electronic and optoelectronic industries is very high in the form of materials such as indium doped tin oxide (ITO), CIS, CIGS, InP, InN, InGaAs, InAlAs, etc., making it one of the most scarce elements in the near future. In this context, a novel sequential PC approach is explored for the fabrication of CIGS thin-films. Deposition of Cu-Ga-Se films is carried out by optimizing the deposition voltage in the first stage followed by subsequent deposition of In in the second stage. The sequentially deposited Cu-Ga-Se/In thin-films are annealed in Ar atmosphere and characterized.

The optimized CuGaSe/In annealed films are noted to have a compact morphology (**Figure 9a**), which is well-suited for application as solar absorber layers, since it facilitates easier diffusion of minority charge carriers and reduces recombination. XRD and Raman analyses confirm the presence of chalcopyrite CIGS films without any undesired phases as shown in **Figure 9b** and **c**. In addition to the micro-Raman analysis of CIGS films, Raman mapping of the optimized films is performed to further verify their phase-purity. **Figure 9d** shows the Raman mapping wherein red refers to the dominant CIGS, green to $Cu_{2-x}Se$ and blue to In_2Se_3 phase. The map of annealed films contains only CIGS phase, thereby indicating the absence of any secondary phases. It also affirms the fact that the elements undergo interdiffusion during the annealing and form the desired chalcopyrite CIGS phase.

TEM analysis is used to confirm the inter-diffusion of In and to infer the CIGS phase formation. The TEM image of CIGS particles is shown in **Figure 10a**. The selected area diffraction (SAED) pattern and TEM-EDS analyses are included as an inset in **Figure 10a**. SAED shows the orientations corresponding to (112), (220), (312) and (400) of CIGS, which corroborate the observation from XRD. TEM-EDS analysis shown in **Figure 10a** confirm the presence of Cu, In Ga and Se affirming the interdiffusion of In (deposited in stage II) during annealing. **Figure 10b** shows the Mott-Schottky ($1/C^2$ vs. V) plot of CIGS thin-films in 0.5 M H_2SO_4 at a frequency of 10 kHz in the dark. The slope of the MS plots is negative, thereby confirming that the CIGS film is p-type. The flat band potential was found to be −0.15 V (vs. SCE) by extrapolating

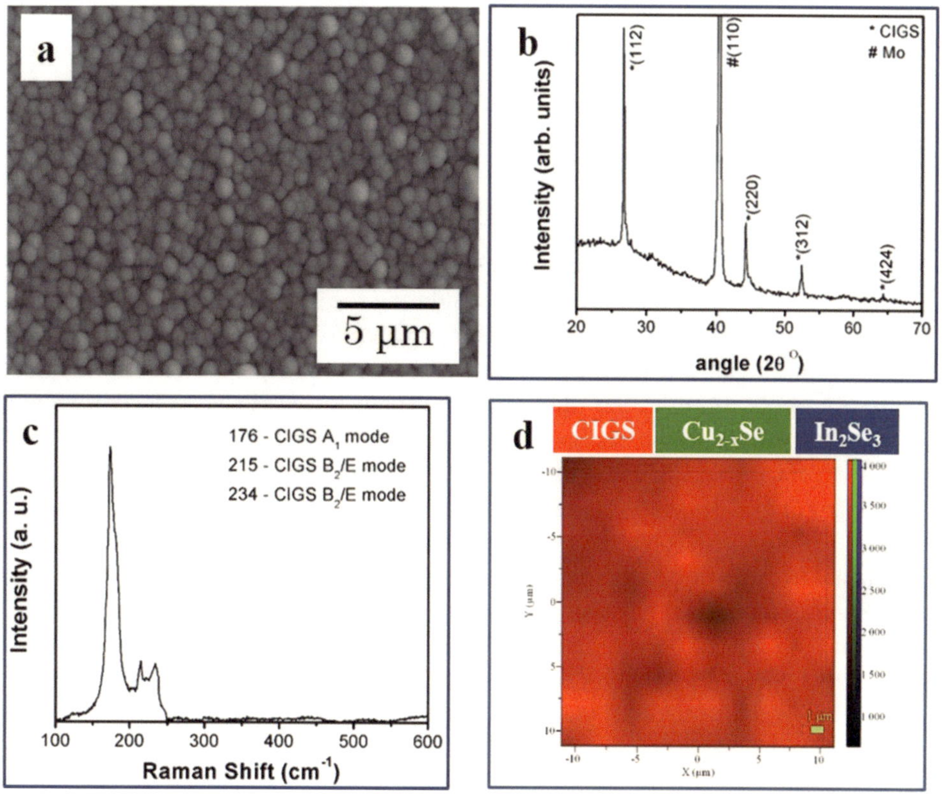

Figure 9. (a) Surface morphology, (b) XRD pattern, (c) Raman spectrum and (d) surface Raman mapping of sequentially pulse electrodeposited CIGS films.

the linear section to x-axis. In addition, $N_a \approx 2.6 \times 10^{16}$ cm^{-3}, calculated using the slope of the curve. The flat-band potential and acceptor density determined herein are close to the values reported previously [46]. **Figure 10c** shows the amperometric current-time (I-t) curve of CIGS films obtained at −0.4 V by chopped light, which demonstrates the nature of photoactivity of

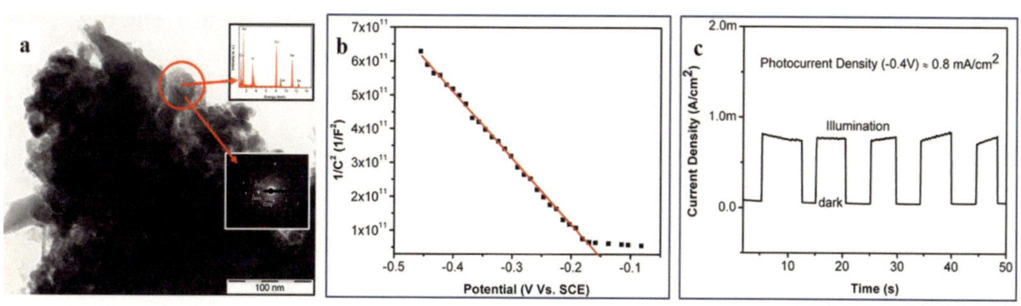

Figure 10. (a) TEM image (inset: TEM-EDS, SAED pattern), (b) Mott-Schottky analysis and (c) amperometric J-t curve of sequentially pulse electrodeposited CIGS films.

CIGS films with a photocurrent density of ≈ 0.8 mA/cm^2. The improved photoresponse of CIGS films indicate their potential for application in thin-film solar cells and photoelectrochemical hydrogen generation.

5. Conclusions

Electrodeposition is a versatile technique for the growth of semiconductor thin films on large areas with a very low capital investment. Considering the state of energy requirement, it is pertinent to explore this low cost technique for production of copper indium selenide (CIS) and copper indium gallium selenide (CIGS) semiconductor films which are the potential candidates for application in solar photovoltaics. The features of electrodeposition with its advanced techniques are discussed in detail while also reviewing the utilization of these for the fabrication of CIS and CIGS thin films. A state-of-the-art summary has been presented on the direct current and pulse electrodeposition of CIS and CIGS thin films detailing various approaches explored while obtaining high efficient CIGS devices. In addition, a new low cost environmental friendly pulse electrodeposition technique has been proposed for the fabrication of CIS and CIGS thin films. In conclusion, the chapter puts forward the idea to photovoltaic community to explore the economic pulse electrodeposition technique for the fabrication of high quality CIGS semiconductor thin films for application in thin films solar cells.

Author details

Sreekanth Mandati[1,2]*, Bulusu V. Sarada[1], Suhash R. Dey[2] and Shrikant V. Joshi[3]

*Address all correspondence to: mandatisree@gmail.com

1 Center for Solar Energy Materials, International Advanced Research Center for Powder Metallurgy and New Materials (ARCI), Hyderabad, Telangana, India

2 Department of Materials Science and Metallurgical Engineering, Indian Institute of Technology Hyderabad, Sangareddy, Telangana, India

3 Department of Subtractive and Additive Manufacturing, University West, Trollhatten, Sweden

References

[1] Green MA, Hishikawa Y, Warta W, Dunlop ED, Levi DH, Hohl-Ebinger J, Ho-Baillie AW. Progress in photovoltaics: Research and applications. Solar Cell Efficiency Tables (version 50). 2017;**25**:668-676

[2] Horig W, Neumann H, Sobotta H, Schumann B, Kohn G. The optical properties of $CuInSe_2$ thin films. Thin Solid Films. 1978;**48**:67-72

[3] Neumann H. Optical properties and electronic band structure of $CuInSe_2$. Solar Cells. 1986;**16**:317-333

[4] Sobotta H, Neumann H, Riede V, Kuhn G, Seltmluvn J, Opperma D. Influence of impurities and free carriers on the optical properties of $CuInSe_2$. Physica Status Solidia. 1980;**60**:531-537

[5] Abou-Elfotouh F, Dunlavy DJ, Cahen D, Noufi IR, Kazmerski LL, Bachmann KJ. Photoluminescence studeies of $CuInSe_2$: Identification of intrinsic defect levels. Progress in Crystal Growth and Characterization. 1984;**10**:15-17

[6] Bardeleben HJV. The chemistry of structural defects in $CuInSe_2$. Solar Cells. 1986;**16**: 381-390

[7] Dagan G, Abou-Elfotouh F, Dunlavy DJ, Matson RJ, Cahen D. Defect level identification in $CuInSe_2$ from photoluminescence studies. Chemistry of Materials. 1990;**2**:286-293

[8] Heyding RD, Murray RM. The crystal structures of $Cu_{1.8}Se$, Cu_3Se_2, α- and γCuSe, $CuSe_2$, and $CuSe_2$II. Canadian Journal of Chemistry. 1976;**54**:841-848

[9] Zahn G, Paufle P. Identification of predominant point defects in nonstoichiometric cuinse by X-ray powder diffraction. Crystal Research and Technology. 1988;**23**:499-507

[10] Albin DS, Tuttle JR, Mooney GD, Carapella JJ, Duda A, Mason A, Noufi R. A study on the optical and microstructural characteristics of quaternary Cu(In.Ga)Se_2 polycrsytalline thin films. In: 21st IEEE Photovoltaics Specialist Conference; 1990. pp. 562-569

[11] Stanbery BJ. Copper indium selenides and related materials for photovoltaic devices. Critical Reviews in Solid State and Materials Sciences. 2002;**27**:73-117

[12] Wei S-H, Zhang SB, Zunger A. Effects of Ga addition to $CuInSe_2$ on its electronic, structural, and defect properties. Applied Physics Letters. 1998;**72**:3199-3201

[13] Abou-Ras D, Kirchartz T, Rau U. Advanced Characterization Techniques for Thin Film Solar Cells. Wiley: Weinheim; 2011. 547 p

[14] Contreras M, Tuttle J, Du D, Qi Y, Swartzlander A, Tennant A, Noufi R. Graded band-gap cu(in,Ga)Se_2 thin-film solar cell absorber with enhanced open-circuit voltage. Applied Physics Letters. 1993;**63**:1824-1826

[15] Miguel A, Contreras BE, Ramanathan K, Hiltner J, Swartzlander A, Hasoon F, Noufi R. Progress toward 20% efficiency in cu(in,Ga)Se_2 polycrystalline thin-film solar cells. Progress in Photovoltaics: Research and Applications. 1999;**7**:311-316

[16] Jackson P, Hariskos D, Wuerz R, Wischmann W, Powalla M. Compositional investigation of potassium doped cu(in,Ga)Se_2 solar cells with efficiencies up to 20.8%. Physica Status Solidi RRL. 2014;**8**:219-222

[17] Philip Jackson DH, Lotter E, Paetel S, Wuerz R, Menner R, Wischmann W, Powalla M. New world record efficiency for cu(in,Ga)Se_2 thin-film solar cells beyond 20%. Progress in Photovoltaics: Research and Applications. 2011;**19**:894

[18] Chirila A, Buecheler S, Pianezzi F, Bloesch P, Gretener C, Uhl AR, Fella C, Kranz L, Perrenoud J, Seyrling S, Verma R, Nishiwaki S, Romanyuk YE, Bilger G, Tiwari AN. Highly

efficient cu(in,Ga)Se_2 solar cells grown on flexible polymer films. Nature Materials. 2011; **10**:857-861

[19] Matson RJ, Jamjoum O, Buonaquisti AD, Russell PE, Kazmerski LL, Sheldon P, Ahrenkiel RK. Metal contacts to $CuInSe_2$. Solar Cells. 1984;**11**:301-305

[20] Broussillou C, Viscogliosi C, Rogee A, Angle S, Grand PP, Bodnar S, Debauche C, Allary JL, Bertrand B, Guillou C, Parissi L, Coletti S. Statistical process control for Cu(In,Ga)$(S,Se)_2$ electrodeposition-based manufacturing process of 60×120cm^2 modules up to 14.0% efficiency. In: IEEE 42nd Photovoltaic Specialist Conference (PVSC); 2015. pp. 1-5

[21] Aksu S, Pethe S, Kleiman-Shwarsctein A, Kundu S, Pinarbasi M. Recent advances in electroplating based CIGS sation. In: 38th IEEE Photovotaics Specialists Conference; 2012. pp. 003092-003097

[22] Bhattacharya RN. CIGS-based solar cells prepared from electrodeposited stacked Cu/In/Ga layers. Solar Energy Materials and Solar Cells. 2013;**113**:96-99

[23] Bhattacharya RN, Hiltner JF, Batchelor W, Contreras MA, Noufi RN, Sites JR. 15.4% $CuIn_{1-x}Ga_xSe_2$-based photovoltaic cells from solution-based precursor films. Thin Solid Films. 2000;**361**:396-399

[24] Dharmadasa IM. Advances in Thin-Film Solar Cells. Singapore: Pan Stanford Publishing; 2013

[25] Lincot D. Electrodeposition of semiconductors. Thin Solid Films. 2005;**487**:40-48

[26] Taunier S, Sicx-Kurdi J, Grand PP, Chomont A, Ramdani O, Parissi L, Panheleux P, Naghavi N, Hubert C, Ben-Farah M, Fauvarque JP, Connolly J, Roussel O, Mogensen P, Mahé E, Guillemoles JF, Lincot D, Kerrec O. Cu(in,Ga)$(S,se)_2$ solar cells and modules by electrodeposition. Thin Solid Films. 2005;**480**:526-531

[27] Lincot JFGD, Taunier S, Guimard D, Sicx-Kurdi J, Chaumont A, Roussel O, Ramdani O, Hubert C, Fauvarque JP, Bodereau N, Parissi L, Panheleux P, Fanouillere P, Naghavi N, Grand PP, Benfarah M, Mogensen P, Kerrec O. Chalcopyrite thin film solar cells by electrodeposition. Solar Energy. 2004;**77**:725-737

[28] Bhattacharya RN. Solution growth and electrodeposited $CuInSe_2$ thin films. Journal of the Electrochemical Society. 1983;**130**:2040-2042

[29] Hibberd CJ, Chassaing E, Liu W, Mitzi DB, Lincot D, Tiwari AN. Non-vacuum methods for formation of cu(in, Ga)$(se, S)_2$ thin film photovoltaic absorbers. Progress in Photovoltaics: Research and Applications. 2010;**18**:434-452

[30] Kemell M, Ritala M, Leskelä M. Thin film deposition methods for $CuInSe_2$ solar cells. Critical Reviews in Solid State and Materials Sciences. 2005;**30**:1-31

[31] Hodes G, Engelhard T, Cahen D, Kazmerski LL, Herrington CR. Electroplated $CuInS_2$ and $CuInSe_2$ layers: Preparation and physical and photovoltaic characterization. Thin Solid Films. 1985;**128**:93-106

[32] Herrero J, Ortega J. Electrodeposition of cu-in alloys for preparing $CuInS_2$ thin films. Solar Energy Materials. 1990;**20**:53-65

[33] Massaccesi S, Sanchez S, Vedel J. Electrodeposition of indium selenide In_2Se_3. Journal of Electroanalytical Chemistry. 1996;**412**:95-101

[34] Bhattacharya RN. Electrodeposited two-layer cu–in–Ga–se/in–se thin films. Journal of the Electrochemical Society. 2010;**157**:D406-D410

[35] Wang X, Wang G, Tian B, Wan S, Du Z. $CuInSe_2$ thin films obtained by pulse-plating electrodeposition technique with novel pulse wave. Chinese Science Bulletin. 2010; **55**:1854-1858

[36] Whang T-J, Hsieh M-T, Kao Y-C, Lee S-J. A study of electrodeposition of $CuInSe_2$ thin films with triethanolamine as the complexing agent. Applied Surface Science. 2009;**255**:4600-4605

[37] Kemell M, Ritala M, Leskela M. Effects of post-deposition treatments on the photoactivity of $CuInSe_2$ thin films deposited by the induced co-deposition mechanism. Journal of Materials Chemistry. 2001;**11**:668-672

[38] Bhattacharya RN, Oh M-K, Kim Y. CIGS-based solar cells prepared from electrodeposited precursor films. Solar Energy Materials and Solar Cells. 2012;**98**:198-202

[39] Fernández AM, Bhattacharya RN. Electrodeposition of $CuIn_{1-x}Ga_xSe_2$ precursor films: Optimization of film composition and morphology. Thin Solid Films. 2005;**474**:10-13

[40] Ugarte R, Schrebler R, CoÂrdova R, Dalchiele EA, GoÂmez H. Electrodeposition of $CuInSe_2$ thin films in a glycine acid medium. Thin Solid Films. 1999;**340**:117-124

[41] Thouin L, Massaccesi S, Sanchez S, Vedel J. Formation of copper indium diselenide by electrodeposition. Journal of Electroanalytical Chemistry. 1994;**374**:81-88

[42] Pottier D, Maurin G. Preparation of polycrystalline thin films of $CuInSe_2$ by electrodeposition. Journal of Applied Electrochemistry. 1989;**19**:361-367

[43] Kampmann A, Sittinger V, Rechid J, Reineke-Koch R. Large area electrodeposition of $cu(in,Ga)Se_2$. Thin Solid Films. 2000;**361-362**:309-313

[44] Bhattacharya RN, Wiesner H, Berens TA, Matson RJ, Keane J, Ramanathan K, Swartzlander A, Mason A, Noufi RN. 12.3% efficient $CuIn_{1-x}Ga_x Se_2$-based device from electrodeposited precursor. Journal of the Electrochemical Society. 1997;**144**:1376-1379

[45] Bhattacharya RN, Batchelor W, Hiltner JF, Sites JR. Thin-film $CuIn_{1-x}Ga_xSe_2$ photovoltaic cells from solution-based precursor layers. Applied Physics Letters. 1999;**75**:1431-1433

[46] Valderrama RC, Sebastian PJ, Enriquez JP, Gamboa SA. Photoelectrochemical characterization of CIGS thin films for hydrogen production. Solar Energy Materials and Solar Cells. 2005;**88**:145-155

[47] Calixto ME, Dobson KD, McCandless BE, Birkmire RW. Controlling growth chemistry and morphology of single-bath electrodeposited $cu(in, Ga)Se_2$ thin films for photovoltaic application. Journal of the Electrochemical Society. 2006;**153**:G521-G528

[48] Long F, Wang WM, Li JJ, Zou ZG. One-step electrodeposition of CIGS thin films from alcohol solution on flexible substrate. Key Engineering Materials. 2008;**368**:472-475

[49] Oda Y, Minemoto T, Takakura H. Electrodeposition of crack-free $CuGaSe_2$ thin films from single bath. Journal of the Electrochemical Society. 2008;**155**:H292-H295

[50] Ribeaucourt L, Savidand G, Lincot D, Chassaing E. Electrochemical study of one-step electrodeposition of copper–indium–gallium alloys in acidic conditions as precursor layers for cu(in,Ga)Se_2 thin film solar cells. Electrochimica Acta. 2011;**56**:6628-6637

[51] Donglin X, Man X, Jianzhuang L, Xiujian Z. Co-electrodeposition and characterization of cu (in, Ga)Se_2 thin films. Journal of Materials Science. 2006;**41**:1875-1878

[52] Ganchev M, Kois J, Kaelin M, Bereznev S, Tzvetkova E, Volobujeva O, Stratieva N, Tiwari A. Preparation of cu(in,Ga)Se_2 layers by selenization of electrodeposited cu–in–Ga precursors. Thin Solid Films. 2006;**511-512**:325-327

[53] Liu J, Liu F, Lai Y, Zhang Z, Li J, Liu YL. Effects of sodium sulfamate on electrodeposition of cu(in,Ga)Se_2 thin film. Journal of Electroanalytical Chemistry. 2011;**651**:191-196

[54] Dergacheva M, Urazov K. Electrodeposition of $CuIn_xGa_{1-x}Se_2$ thin films from sulfosalicylic acid. Electrochimica Acta. 2013;**107**:120-125

[55] Chraibi F, Fahoume M, Ennaoui A, Delplancke JL. Influence of citrate ions as complexing agent for electrodeposition of $CuInSe_2$ thin films. Physica Status Solidi (a). 2001;**186**: 373-381

[56] Mishra KK, Rajeshwar K. A voltammetric study of the electrodeposition chemistry in the cu + in + se system. Journal of Electroanalytical Chemistry and Interfacial Electrochemistry. 1989;**271**:279-294

[57] Zhang L, Jiang FD, Feng JY. Formation of $CuInSe_2$ and cu(in,Ga)Se_2 films by electrodeposition and vacuum annealing treatment. Solar Energy Materials and Solar Cells. 2003;**80**:483-490

[58] Liu F, Huang C, Lai Y, Zhang Z, Li J, Liu Y. Preparation of cu(in,Ga)Se_2 thin films by pulse electrodeposition. Journal of Alloys and Compounds. 2011;**509**:L129-L133

[59] Bouabid K, Ihlal A, Manar A, Outzourhit A, Ameziane EL. Effect of deposition and annealing parameters on the properties of electrodeposited $CuIn_{1-x}Ga_xSe_2$ thin films. Thin Solid Films. 2005;**488**:62-67

[60] Friedfeld R, Raffaelle RP, Mantovani JG. Electrodeposition of $CuIn_xGa_{1-x}Se_2$ thin films. Solar Energy Materials and Solar Cells. 1999;**58**:375-385

[61] Kang F, Ao J, Sun G, He Q, Sun Y. Structure and photovoltaic characteristics of $CuInSe_2$ thin films prepared by pulse-reverse electrodeposition and selenization process. Journal of Alloys and Compounds. 2009;**478**:L25-L27

[62] Caballero-Briones F, Palacios-Padrós A, Sanz F. $CuInSe_2$ films prepared by three step pulsed electrodeposition. Deposition mechanisms, optical and photoelectrochemical studies. Electrochimica Acta. 2011;**56**:9556-9567

[63] Valdés MH, Vázquez M. Pulsed electrodeposition of p-type $CuInSe_2$ thin films. Electrochimica Acta. 2011;**56**:6866-6873

[64] Chitra V, Vasantha S, Murali KR. Electrochemical phototvoltaic cells with pulse electrodeposited copper indium selenide films. In: 2011 International Conference on Nanoscience, Engineering and Technology (ICONSET). pp. 429-432

[65] Hu S-Y, Lee W-H, Chang S-C, Cheng Y-L, Wang Y-L. Pulsed electrodeposition of $CuInSe_2$ thin films on to Mo-glass substrates. Journal of the Electrochemical Society. 2011;**158**: B557-B561

[66] Prasher D, Rajaram P. Growth and characterization of pulse electrodeposited $CuInSe_2$ thin films. Electronic Materials Letters. 2012;**8**:515-518

[67] Valdés M, Vázquez M. Composition, morphology, and optical properties of CuInSe2 thin films electrodeposited using constant and pulsed potentials. Journal of Solid State Electrochemistry. 2012;**16**:3825-3835

[68] Shanmugavel A, Srinivasan K, Murali KR. Pulse electrodeposited copper indium sulpho selenide films and their properties. Materials Science in Semiconductor Processing. 2013;**16**:1665-1671

[69] Murali K. Optical properties of pulse plated $CuInSe_2$ thin films. International Journal of Engineering and Innovative Technology. 2014;**3**:207-211

[70] Fu Y-P, You R-W, Lew KK. $CuIn_{1-x}Ga_xSe_2$ absorber layer fabricated by pulse-reverse electrodeposition technique for thin films solar cell. Journal of the Electrochemical Society. 2009;**156**:D553-D557

[71] Bi J, Yao L, Ao J, Gao S, Sun G, He Q, Zhou Z, Sun Y, Zhang Y. Pulse electro-deposition of copper on molybdenum for cu(in,Ga)Se_2 and $Cu_2ZnSnSe_4$ solar cell applications. Journal of Power Sources. 2016;**326**:211-219

[72] Bi J, Ao J, Gao Q, Zhang Z, Sun G, He Q, Zhou Z, Sun Y, Zhang Y. Controllable growth of Ga film electrodeposited from aqueous solution and cu(in,Ga)Se_2 solar cells. ACS Applied Materials & Interfaces. 2017;**9**:18682-18690

[73] Mandati S, Sarada BV, Dey SR, Joshi SV. Photoelectrochemistry of cu(in, Ga) se_2 thin-films fabricated by sequential pulsed electrodeposition. Journal of Power Sources. 2015; **273**:149-157

[74] Mandati S. Fabrication of $CuInSe_2$ and cu (in, Ga) Se_2 absorber layers by pulse-and pulse-reverse electrochemical techniques for solar photovoltaic applications. Physical Review D. 2015;**82**:1-49

[75] Mandati S, Sarada BV, Dey SR, Joshi SV. Improved photoelectrochemical performance of Cu(In,Ga)Se_2 thin films prepared by pulsed electrodeposition. Journal of Renewable and Sustainable Energy. 2013;**5**:031602-1-031602-7

[76] Mandati S, Sarada BV, Dey SR, Joshi SV. Pulsed electrodeposition of $CuInSe_2$ thin films with morphology for solar cell applications. Journal of the Electrochemical Society. 2013;**160**:D173-D177

[77] Chandrasekar MS, Pushpavanam M. Pulse and pulse reverse plating—Conceptual, advantages and applications. Electrochimica Acta. 2008;**53**:3313-3322

[78] Dharmadasa IM, Burton RP, Simmonds M. Electrodeposition of $CuInSe_2$ layers using a two-electrode system for applications in multi-layer graded bandgap solar cells. Solar Energy Materials and Solar Cells. 2006;**90:**2191-2200

[79] Zhou Y-L, Zhou W-H, Li M, Du Y-F, Wu S-X. Hierarchical Cu_2ZnSnS_4 particles for a low-cost solar cell: Morphology control and growth mechanism. The Journal of Physical Chemistry. 2011;**C115**:19632-19639

[80] Neumann H, Tomlinson RD, Nowak E, Avgerinos N. Electrical properties of p-vpe $CuInSe_2$ single crystals. Physica Status Solidi A. 1979;**56**:K137-K140

[81] Wada T, Kohara N, Nishiwaki S, Negami T. Characterization of the cu(in,Ga)Se_2 Mo interface in CIGS solar cells. Thin Solid Films. 2001;**387**:118-122

[82] Duchatelet A, Sidali T, Loones N, Savidand G, Chassaing E, Lincot D. 12.4% efficient Cu(In,Ga)Se_2 solar cell prepared from one step electrodeposited Cu–In–Ga oxide precursor layer. Solar Energy Materials and Solar Cells. 2013;**119**:241-245

[83] Duchatelet A, Letty E, Jaime-Ferrer S, Grand PP, Mollica F, Naghavi N. The impact of reducing the thickness of electrodeposited stacked cu/in/Ga layers on the performance of CIGS solar cells. Solar Energy Materials and Solar Cells. 2017;**162**:114-119

[84] Insignares-Cuello C, Oliva F, Neuschitzer M, Fontané X, Broussillou C, de Monsabert TG, Saucedo E, Ruiz C, Pérez-Rodríguez A, Izquierdo-Roca V. Advanced characterization of electrodeposition-based high efficiency solar cells: Non-destructive Raman scattering quantitative assessment of the anion chemical composition in cu (in, Ga)(S, se)$_2$ absorbers. Solar Energy Materials and Solar Cells. 2015;**143**:212-217

[85] Romanyuk YE, Hagendorfer H, Stücheli P, Fuchs P, Uhl AR, Sutter-Fella CM, Werner M, Haass S, Stückelberger J, Broussillou C, Grand P-P, Bermudez V, Tiwari AN. All solution-processed chalcogenide solar cells—From single functional layers towards a 13.8% efficient CIGS device. Advanced Functional Materials. 2015;**25**:12-27

[86] Ye H, Park HS, Akhavan VA, Goodfellow BW, Panthani MG, Korgel BA, Bard AJ. Photoelectrochemical characterization of $CuInSe_2$ and cu($In_{1-x}Ga_x$)Se_2 thin films for solar cells. The Journal of Physical Chemistry C. 2011;**115**:234-240

[87] Jadhav HS, Kalubarme RS, Ahn S, Yun JH, Park C-J. Effects of duty cycle on properties of CIGS thin films fabricated by pulse-reverse electrodeposition technique. Applied Surface Science. 2013;**268**:391-396

[88] Vadivel S, Srinivasan K, Murali KR. Structural and optoelectronic properties of pulse plated $CuIn_{1-x}Ga_xS_2$ films. Optik: International Journal for Light and Electron Optics. 2015;**126**:558-563

Chapter 7

The Electrochemical Performance of Deposited Manganese Oxide-Based Film as Electrode Material for Electrochemical Capacitor Application

Chan Pei Yi and Siti Rohana Majid

Additional information is available at the end of the chapter

http://dx.doi.org/10.5772/intechopen.71957

Abstract

The transition metal oxide has been recognized as one of the promising electrode materials for electrochemical capacitor application. Due to the participation of charge transfer reactions, the capacitance offered by transition metal oxide can be higher compared to double layer capacitance. The investigation on hydrous ruthenium oxide has revealed the surface redox reactions that contributed to the wide potential window shown on cyclic voltammetry curve. Although the performance of ruthenium oxide is impressive, its toxicity has limited itself from commercial application. Manganese oxide is a pseudocapacitive material behaves similar to ruthenium oxide. It consists of various oxidation states which allow the occurrence of redox reactions. It is also environmental friendly, low cost, and natural abundant. The charge storage of manganese oxide film takes into account of the redox reactions between Mn^{3+} and Mn^{4+} and can be accounted to two mechanisms. The first one involves the intercalation/deintercalation of electrolyte ions and/or protons upon reduction/oxidation processes. The second contributor for the charge storage is due to the surface adsorption of electrolyte ions on the electrode surface.

Keywords: pseudocapacitors, energy storage devices, metal oxides, thin films, electrode materials

1. Introduction

The electronics technologies we are granted nowadays are the results of many years' researches. From Benjamin Franklin, Alessandro Volta, Michael Faraday, and Nikola Tesla, the continuous efforts have developed the electrical knowledge which has been practicalized for the sake of living standards. The better quality of life and the transformation from agricultural-based economy into information-based economy indicating the globalization has taken place. Although the fruits of globalization are attractive, the globalization also causes unavoidable negative consequence such as global warming. The increased concentration of greenhouse gases in the atmosphere is referred as the impact of human activities such as the burning of fossil fuel for transports and power electricity plants. It brings into our concern that an alternative energy source is required, at the same time solving the energy demand issue that caused by globalization is needed. It is therefore, necessary to improve the energy management. Energy storage system, as part of the energy management, receives tremendous attentions for this purpose.

Energy storage system is a big family comprising various energy storage devices, for example, solar cell, battery, electrochemical capacitor, and fuel cell, which can be used for specific applications, for instance, the employment of fuel cell-based system over gasoline engine in transportation [1]. By reducing the rated fuel cell power and controlling the energy loss, the fuel efficiency can be enhanced [2]. Another technique of energy management is by distributing the power onto different energy sources. For example, a hybrid system based on a polymer electrolyte membrane fuel cell and nickel-metal hydride battery has been evaluated for tramway in Spain [3]. Tsukahara and Kondo have inspected the prospective hybridizations of fuel cell with Li-ion battery and electric double layer capacitor (EDLC) to power the railway vehicles [4].

Apart from transportation, many portable electronic devices also supported by the battery, that is also an energy storage device. From nickel-cadmium battery, nickel-metal hydride battery, to Li-ion battery, the studies on batteries have never stopped. To date, there are more than 35,000 published papers regarding to lithium battery according to Web of Science. Nowadays, Li-ion battery is the most promising battery. A diagram describing charging-discharging mechanism of Li-ion battery is displayed in **Figure 1**.

As a rechargeable battery, Li-ion battery encourages a rapid growth in the mobile devices technology. Abraham has outlined the limitation and future outlook of Li-ion battery [5]. Li-O_2 or Li-air battery will be the battery that can achieve the utmost energy density. However, a down-to-earth Li-air battery is still well on the way to be practical. Recently, an all-solid-state rechargeable battery based on a solid glass electrolyte promises a safe, low cost, and longer cycle life properties for battery [6]. The performances of batteries are summarized in Ragone plot [7], **Figure 2**.

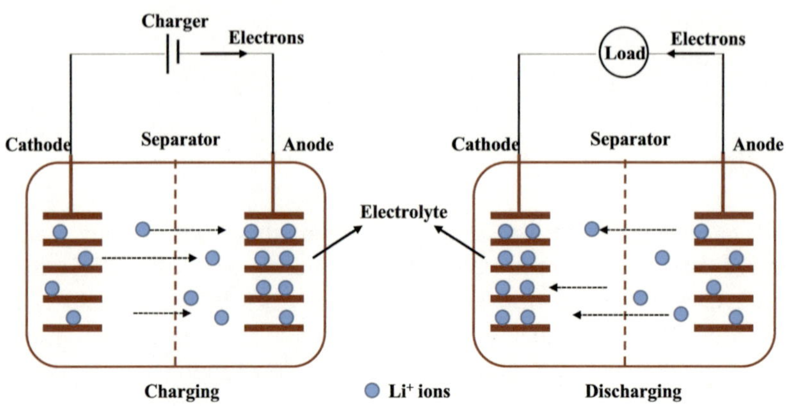

Figure 1. Charging-discharging schematic diagram of li-ion rechargeable battery.

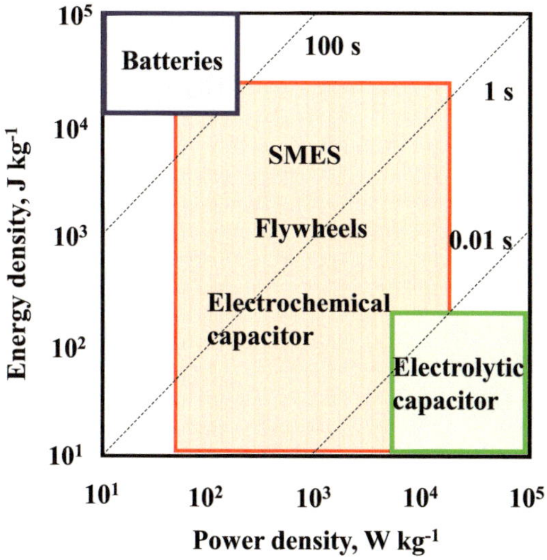

Figure 2. Ragone plot.

Nowadays, the Ragone plot, which is an indication of possible development for certain energy storage devices, has included some storage devices other than batteries. For example, fuel cell, capacitor, and supercapacitor. However, it does not show all the other important properties for instance cycle stability, temperature range of operation, and energy efficiency. Therefore, the Ragone plot cannot be the only reference for evaluating the performance of an energy storage device. Nevertheless, it can be used as a source of information which is continually updated and improved [8].

An electrochemical capacitor, also known as supercapacitor, is an energy storage device whereby its electrochemical performance lies in between a conventional capacitor and a battery. It can store larger amount of energy compared to a capacitor, but lesser when compared to a battery. Nonetheless, an electrochemical capacitor has higher power density than a battery. The electrochemical capacitor is categorized into two types: electric double layer capacitor (EDLC) and pseudocapacitor. The working principle behind an EDLC is based on the electrostatic interaction. The EDLC charges/discharges are according to the adsorption/de-adsorption processes. The schematic diagram of an EDLC is depicted in **Figure 3**.

As shown, the positive and negative charges are attracted to the electrode surfaces with opposite charges respectively. The charges are thus stored by means of the electric double layer formed at the electrode surface. Therefore, material with high surface area and conductivity such as carbon materials are suitable to be employed as electrode for EDLC. Conversely, a pseudocapacitor makes use of the charge transfer reactions for charge storage purpose. The corresponding capacitance is named as pseudocapacitance.

Although the charge storage mechanisms are different, the electrochemical signatures of EDLC and pseudocapacitor are similar when they are examined using cyclic voltammetry test [9]. As

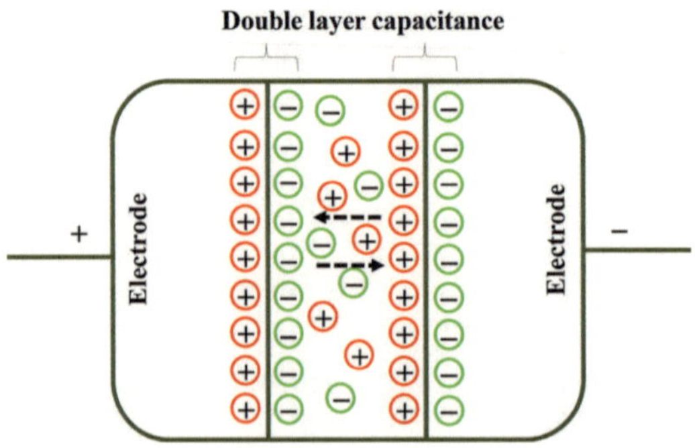

Figure 3. Schematic diagram of EDLC.

the charge stored is linearly dependent with the potential, the resultant voltammetry curve exhibits a rectangular and symmetry shape. The redox materials such as conducting polymers and transition metal oxides are usually studied as the potential electrode material for pseudocapacitor. The hydrous ruthenium oxide is found to exhibit a high specific capacitance of 720 F g^{-1}, making it a promising electrode material for electrochemical capacitor [10]. Nevertheless, ruthenium oxide is expensive and toxic. This has prevented hydrous ruthenium oxide from commercial use. Besides, the usage of strong acidic electrolyte also forbids it from application. Other materials such as SnO_2, MnO_2, TiO_2, VO_2, and MoO_3 have been studied as the alternative electrode material and they are showing different potentials for the practical applications [11–15]. Among the materials, manganese oxide has been widely studied due to the fact that the manganese oxide can form a variety of composites with different materials using various synthesis routes [16–18]. In the following section, we will discuss about the properties of manganese oxide.

2. Manganese oxide

Manganese is one of the most abundant elements on Earth which is widely distributed across the crust. In general, manganese deposit forms from seawater. Thus, its redox sensitivity toward oxidation state of ocean enables us to explore the ancient environmental conditions [19]. In human body, the manganese participates in carbohydrate metabolism as well as formation of bone and connective tissues. Manganese oxide has been used since ancient time. It was employed for elucidation of glass and acted as black pigment. Throughout the years, the researches on manganese oxide have deepen the understanding on the chemistry of this compound [20–26]. At the same time, the utilization of manganese oxide has been exploited and diversified.

The advantage of manganese oxide lies in the feasible formation of various structures based on different arrangements of basic building structure of manganese oxide, which is MnO_6 octahedral. This is a structure where O^{2-} ions are octahedrically coordinated to the central of Mn^{4+} ion, as shown in **Figure 4**.

http://dx.doi.org/10.5772/intechopen.71957

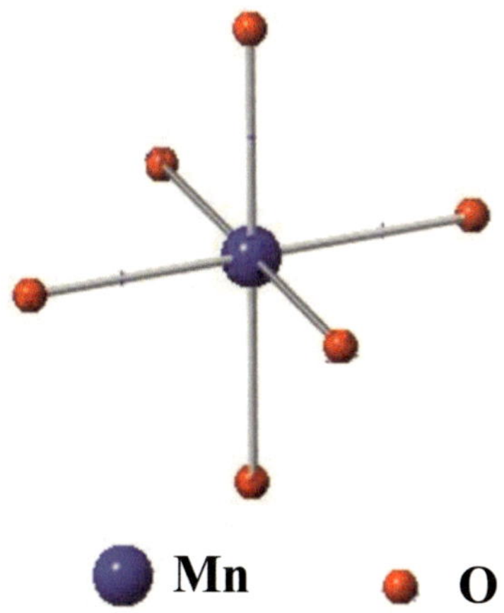

Figure 4. Schematic diagram of MnO_6 octahedral.

The octahedral structure can be arranged through edge- and/or corner-sharing. In common, there are two types of structure: (a) tunnel or chain structure and (b) layer structure. Chain structure is made up of corner-sharing arrangement while tunnel structure is resulted from the combination of single, double, or triple chains of MnO_6 octahedral. Layer structure is constructed by the sheets or layers stacking of MnO_6 octahedral. Both kinds of structure can hold proton and electrolyte cation.

As a transition metal, manganese exists in various valence states which in turns form a variety of mineral with distinct physical and chemical properties. When different phases of manganese oxide are mixed, they can intergrow to form a new structure. The manganese oxides mineral is inclusive of MnO_2, Mn_2O_3, Mn_3O_4, and MnOOH. The different forms of minerals are named as the polymorphs. For instance, MnO_2 has three polymorphs: pyrolusite, ramsdellite, and nautile while MnOOH has polymorphs of manganite, groutite, and feiknechtite. The polymorphic form is dependent on the linkage between MnO_6 octahedrals, which is determined by the preparation process of manganese oxide [27]. Over the years, the variation in structural forms and properties has made manganese oxide a suitable candidate for different applications such as energy storage system, biosensor, coating, environmental, and nuclear science [28–34]. According to Pourbaix diagram, **Figure 5**, the electrochemical performances of manganese oxides in aqueous media are arisen from its various oxidation states (Mn^{2+}, Mn^{3+}, and Mn^{4+}) that are emerged from different oxide phases (Mn_3O_4, Mn_2O_3, and MnO_2) which are thermodynamically stable in alkaline region [35].

In general, the charge storage of manganese oxide relies on two mechanisms: (a) surface adsorption of electrolyte ions and (b) intercalation/deintercalation of electrolyte ions and/or proton upon reduction/oxidation [36]. Although higher surface area can lead to higher value of specific capacitance, further increasing in surface area is found to not contribute

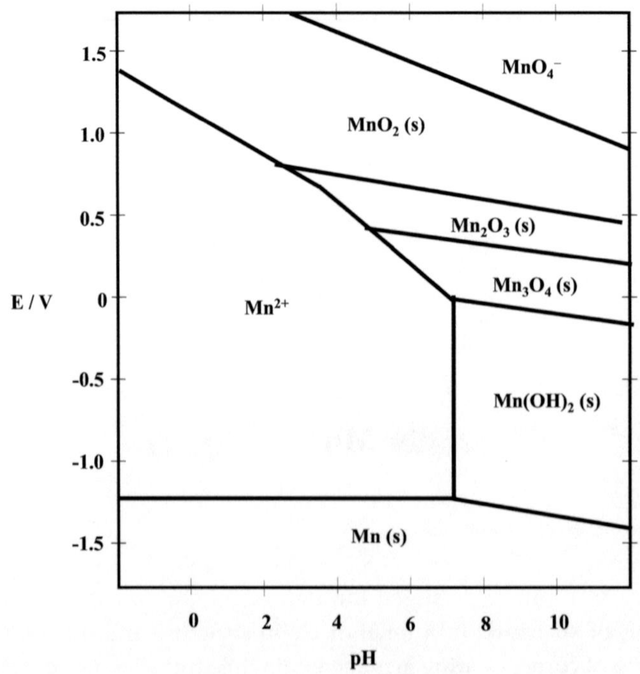

Figure 5. Pourbaix diagram of manganese.

to the specific capacitance [12, 37]. The structure of the manganese oxide can determine its electrochemical performance. A structure with more rooms for the insertion of electrolyte ions will offer higher charge storage capacity, and subsequently higher specific capacitance. To date, the researchers are still working best to figure out the charge storage mechanism of manganese oxide in order to better explain its electrochemical behavior [38–40]. This knowledge will inspire us and provide us a way on how to fully utilize the potential capacitance of manganese oxide.

Manganese oxide-based film can be prepared via a variety of methods. A material will exhibit different properties such as particle sizes and types of defects, depending on the fabrication routes. In addition to this, the change in experimental parameters also produces materials with different electrochemical properties. The common preparation methods are the hydrothermal, chemical bath deposition, polyol synthesis, sol–gel, electrodeposition, solvothermal, and co-precipitation [20, 36, 39, 41–45]. The specific capacitances obtained ranges from 121.5 to 698 F g^{-1}, which is still lower than the theoretical specific capacitance of manganese oxide (1380 F g^{-1}). To better utilize the electrochemical active sites, one should understand the relationship between the experimental method and the corresponding structure formed. This chapter will focus on the green and relatively simple method of electrodeposition. Thereafter, we will discuss about the electrochemical performance of manganese oxide-based film fabricated using electrodeposition technique.

http://dx.doi.org/10.5772/intechopen.71957

3. General principle of electrodeposition

Electrodeposition refers to an electrical process such as electrolytic and electrophoretic deposition, which allows the accumulated mass of a metal ions, or deposit, coated onto an electrode. A common configuration of electrodeposition is displayed in **Figure 6**.

It is usually made up of a working electrode, counter electrode, and reference electrode. The working electrode is the substrate where the deposition reaction takes place. The reference electrode is used to maintain the voltage stability for the working electrode while the counter electrode or auxiliary electrode is utilized to complete the current flow. There is various reference electrodes served for different electrolyte solutions. A standard hydrogen electrode (SHE) consists of 1.0 M H^+ (aq.) solution and is used to compare with other reference electrodes since the standard electrode potential of hydrogen is 0 V. Saturated calomel electrode (SCE) is a reference electrode composes of KCl solution and establishes based on the reaction between elemental mercury and mercury chloride. However, the dangerous nature of mercury content has prohibited the SCE reference electrode from widely use. Instead, the silver chloride (Ag/AgCl) electrode is employed. Ag/AgCl reference electrode is mostly utilized for electrochemical tests and industrial application due its simplicity for fabrication, stability, and non-toxicity. Nonetheless, the usage of Ag/AgCl in basic solution must be heedful because a long immersion of electrode allows the diffusion of OH^- ions into the internal filling solution, which can subsequently increase the pH and shift the reaction potential. The measurement of potential using various reference electrodes can be referred to SHE and the corresponding values are recorded in **Figure 7**.

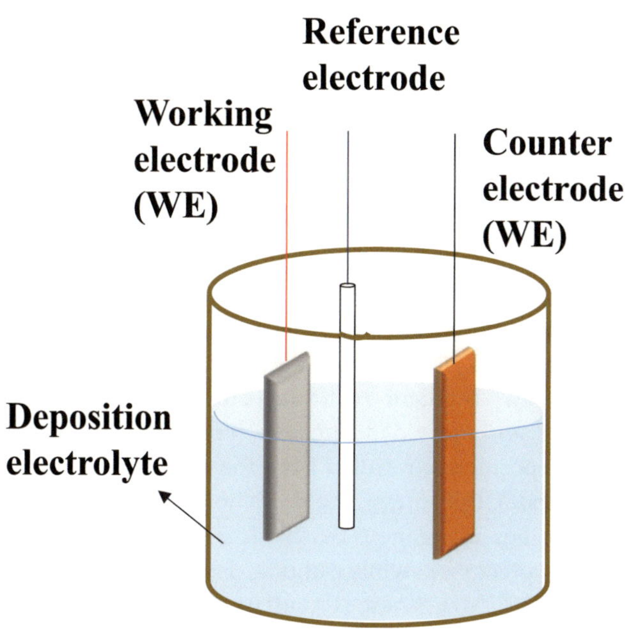

Figure 6. General electrodeposition setup.

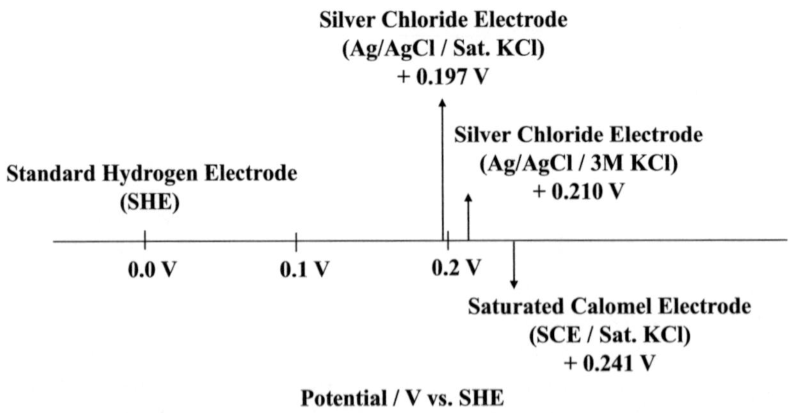

Figure 7. Voltage conversion between reference electrodes.

As electrolytic deposition makes use of solution containing metal ions for deposition purpose, electrophoretic deposition utilizes a mixture consists of suspended colloidal particles. Electrophoretic deposition was revealed when a Russian scientist observed the movement of clay particles in water induced by electric field. It occurs when the non-conductive electrically charged particles migrate to the electrode surface under an applied electric field. The charged particles suspend in the electrolyte due to the mutual electrostatic repulsion [46]. Inside the suspension, the surface charges attract the electrolyte ions with opposite charge. As a result, the counter ions adsorb onto the surface charges, forming a diffuse cloud of counter ions. The process is governed by electrostatic interaction. At the same time, the adsorbed ions repulse from each other. When they are close enough to overcome the electrostatic force, van der Waals attraction will be predominant and adhesion can occur. To achieve electrophoretic deposition, a stable suspension containing well-dispersed particles with desired electrophoretic mobility must be prepared firstly. The usage of additives such as dopamine and triethanolamine can help to stabilize the suspension [47–48]. In addition, it offers a more uniform and adherent deposit. The parameters that play the role on electrophoretic deposition are composition of dispersion medium, pH of electrolyte, and concentrations of particles and electrolyte. The principles behind electrolytic deposition and electrophoretic deposition have been studied [49–51]. An understanding about these two processes is required in order to fabricate desired electrode material. The oxidation kinetics of manganese oxide from Mn^{2+} is reliant on the deposition methods. Depends on the particle charge, electrodeposition can be divided into two types: anodic deposition and cathodic deposition. The anodic deposition is resulted from the oxidation of negative ions on anode (positively charged electrode). In contrast, the reduction of positive ions on cathode results in a cathodic deposition. For manganese oxide film, the anodic deposition involves the oxidation of cationic Mn^{2+} precursors while cathodic deposition is achieved by the reduction of anionic Mn^{7+} from MnO_4^{-}. Manganese (II) sulfate ($MnSO_4$), manganese (II) nitrate (Mn ($NO_3)_2$), manganese (II) acetate ($Mn(CH_3COO)_2$), and manganese (II) chloride ($MnCl_2$) can be used as precursors for Mn^{2+}. As the deposition mechanism is independent of precursors, it does not affect significantly on the capacitive behaviors of manganese oxides formed [52].

Due to this reason, the mass loading plays an important role in determining the capacity of manganese oxide formed. Among these precursors, $Mn(CH_3COO)_2$ decomposes at lower potential and offers higher deposition rate which make it a favorable precursor. In addition, the Mn^{3+} is more stable with acetate compared to sulfate which makes it a thermodynamic favor for oxidation kinetics [53]. On the other hand, $KMnO_4$ is usually the Mn^{7+} precursor for cathodic deposition.

As the oxidation process can lead to the dissolution of metal substrate, the cathodic reduction is usually preferred over anodic oxidation as various metals can be employed as the substrates [51]. The anodic oxidation of Mn^{2+} involves other oxidation state of manganese ions. Initially the Mn^{2+} will diffuse and adsorb onto the electrode surface to form $Mn_{ads}{}^{2+}$. The adsorbed ions oxidize to Mn^{3+} which subsequently forms intermediate with water. Under appropriate heating temperature, the MnOOH intermediate can be transformed to MnO_2. The detailed oxidation reaction of Mn^{2+} to MnO_2 in the medium with different acidity is described in **Section 4** [54, 55]. According to Pourbaix diagram, the reduction process of Mn^{7+} occurs in a neutral aqueous solution is as follow, **Reaction 1** [35]:

$$MnO_4^- + 2H_2O + 3e^- \rightarrow MnO_2 + 4OH^- \quad (1)$$

The electrodeposition can be carried out either by a constant potential or current technique. For constant potential technique, the resulted current-time transient can be determined by following factors: (1) the potential and time of the oxide formation, (2) the potential and time of oxide reduction, and (3) the maintenance time of electrode at reversible potential before reduction takes place [56]. The chronopotentiometry is an electrodeposition technique performed with a constant current. The constant current is applied between working and auxiliary electrodes while the potential of working electrode is measured against a reference electrode. The potential reveals the reaction takes place on the electrode during the electrodeposition process. Before the electrodeposition starts, there is no concentration gradient of oxidants in the solution regardless of the distance from electrode surface. This initial potential is also called as open circuit potential. However, a concentration gradient of oxidants is formed when the reduction initiates by the application of current. The reduction is a resultant process of oxidants responds to the applied current. As a result, the oxidants near to the electrode surface are consumed, causing the oxidants diffuse from bulk solution toward electrode surface in order to accommodate the reduction process. The potential is comparable to the redox potential of certain electron transfer reaction. Since the concentration changes with time, this potential alters correspondingly. Once the current cannot be further sustained by this redox potential, which is due to the concentration of oxidants turns to zero, this potential will adjust to another redox potential in order to maintain the fixed current. The changing potential during galvanostatic deposition can lead to the formation of various morphologies. Different from chronoamperometry, which is a deposition method employing a constant voltage, a desired morphology can be formed accordingly. Knowledge of the interactions between electrode and electrodeposition parameters allows us to construct and improve the electrode film performance. Herein, the manganese oxide deposited using various deposition techniques are discussed and evaluated.

4. Electrochemical performance of deposited manganese oxide-based film

4.1. Effect of electrolyte composition

For anodic deposition, manganese acetate ($Mn(CH_3COO)_2$) and manganese sulfate ($MnSO_4$) are always chosen as the Mn^{2+} precursors. They offer different kinds of morphologies although the electrodeposition is carried out under the same conditions. For example, the morphology produced from 0.01 M $Mn(CH_3COO)_2$ at constant current density of 30 mA cm^{-2} exhibits interconnected but non-continuous nanorods structure, **Figure 8(a)** [57]. On the other hand, 0.01 M $MnSO_4$ leads to a continuous and homogenous nanorods structure (**Figure 8(b)**).

With the addition of H_2SO_4, a discrete crystallite of manganese oxide is formed ([58]). H_2SO_4 acts as a supporting electrolyte and enhances the stability of the soluble Mn^{3+} intermediate before further reactions [59]. This allows more manganese ions to be deposited and form a film on the substrate after certain electrodeposition time. The electrodeposition mechanism of manganese oxide with the presence of H_2SO_4 is shown as following [60]:

$$Mn^{2+} \rightarrow Mn^{3+} + e^- \tag{2}$$

Disproportionation pathway:

$$2Mn^{3+} \rightarrow Mn^{2+} + Mn^{4+} \tag{3}$$

$$Mn^{4+} + 2H_2O \rightarrow MnO_2 + 4H^+ \tag{4}$$

Hydrolysis pathway:

$$Mn^{3+} + 2H_2O \rightarrow MnOOH + 3H^+ \tag{5}$$

$$MnOOH \rightarrow MnO_2 + H^+ + e^- \tag{6}$$

Depending on the concentration of acid, there are two proposed pathways for the formation of manganese oxide: disproportionation and hydrolysis. **Reaction 2** shows the oxidation of Mn^{2+} become soluble Mn^{3+} intermediate. If the supporting acid, which is H_2SO_4 in this case, has higher

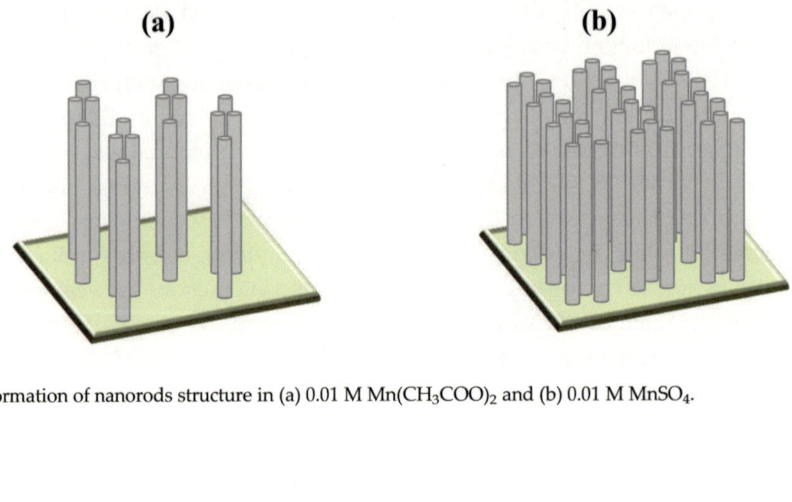

Figure 8. Formation of nanorods structure in (a) 0.01 M $Mn(CH_3COO)_2$ and (b) 0.01 M $MnSO_4$.

concentration, Mn^{3+} achieves relatively higher stability allowing it to experience disproportionation which subsequently forms Mn^{2+} and Mn^{4+} (**Reaction 3**). Mn^{4+} is then hydrolyzed to form MnO_2 on the surface of substrate (**Reaction 4**). On the other hand, at lower acidity condition, Mn^{3+} is less stable and thus can hydrolyze easily to form MnOOH (**Reaction 5**). MnOOH is then converted to MnO_2 under suitable annealing temperature (**Reaction 6**). The manganese oxide electrodeposited from the electrolyte containing H_2SO_4 can obtain specific capacitance as high as 5600 F g^{-1} [59]. A less acidic medium also can be achieved by adding sodium sulfate (Na_2SO_4). It can act as the supporting electrolyte for Mn^{2+} and Mn^{7+} precursors [61, 62].

4.2. Effect of deposition potential

In an electrolyte of $Mn(CH_3COO)_2$, 0.5 V (*vs.* SCE) of anodic potential forms a manganese oxide consists of two oxidation states: trivalent (Mn^{3+}) and tetravalent states (Mn^{4+}), while divalent ions Mn^{2+} and Mn^{3+} are observed for manganese oxide formed at anodic potential lower than 0.5 V [63, 64]. The manganese oxide resulted from deposition potential of 0.2 V (*vs.* SCE) for 1 hour consists of γ-Mn_2O_3 and/or Mn_3O_4 while the phase transforms from tetragonal to hexagonal system when the deposition potential increased to 0.4 V and higher potentials. At anodic potential higher than 0.5 V, Mn^{4+} state becomes dominant and hydrous MnO_2 can be formed [65].

The anodic potential is found to be correlated to the adsorbed water content in the manganese oxide structure. The increase in anodic deposition potential leads to the formation of higher oxidation state of manganese ions, at the same time results in the reduction in adsorbed water content [64, 66]. As the rate of nucleus formation increases more rapidly than the rate of crystal growth with the increase of anodic potential, the morphology preferably grows horizontally, which subsequently forms a compact and layered structure. The formation of this structure is resulted from the local fluctuation of reactants. As the reactant consumption rate increases with higher deposition potential, the reactions take place at the vicinity of electrode also occur rapidly. The release of adsorbed water molecules further resulted in the formation of an uneven surface. The highest specific capacitance obtained is 240 F g^{-1} at 5 mV s^{-1}. A similar specific capacitance value is achieved when the technique applied is cathodic potentiostatic electrodeposition. The MnO_2 prepared through cathodic reduction process at a constant potential of 0.55 V (*vs.* SCE) has achieved around 233 F g^{-1} [67]. However, instead of a smooth surface which is produced at 0.5 V (*vs.* SCE) of anodic potential, a rough surface is formed for this cathodic potentiostatic electrodeposition. A further decrease in deposition potential (−1.8 V) can result in a porous structure with nanowall architecture [68].

For cathodic deposition, the increase in deposition potential leads to the uniform distribution of manganese ions onto the substrate. At relatively low cathodic deposition potential, the morphology formed is a cluster-like structure. The increase of cathodic deposition potential from −0.4 V to −0.1 V motivates the formation of flaky-like structure [62]. Further increase in cathodic deposition potential forms a more homogenous and flat structure. The formation of various structures with the cathodic deposition potential arises from the different deposition mechanisms with different potential values. The first mechanism takes place in the cathodic potential range of 0.1–0.4 V whereby the MnO_4^{2-} produced will form $H_2MnO_4^-$ that subsequently reduce to form manganese oxide, as shown in **Reaction 7**. Another mechanism occurs at cathodic potential less than 0.2 V. It takes into account of the dissolution of manganese oxide that has formed initially. This brings about the formation of Mn^{2+} which is then reacted with MnO_4^- to produce manganese oxide, as described in **Reactions 8** and **9**.

$$3Mn^{2+} + 2MnO_4^- + 2H_2O \rightarrow 5MnO_2 + 4H^+ \quad (7)$$

$$3MnO_4^{3-} + 8H^+ \leftrightarrow MnO_4^- + 2MnO_2 + 4H_2O \quad (8)$$

$$H_2MnO_4^- + e^- \rightarrow 2MnO_2 + 4OH^- \quad (9)$$

Depending on the potential range, a potentiodynamic deposition can produce a nanorod-like structure with various diameters, given that a same scan rate is employed, as shown in **Figure 9**. Smaller potential range (0.1–0.4 V *vs.* SCE) leads to the formation of nanorods structure with higher diameter while wider potential range (0.1–0.6 V *vs.* SCE) tends to form a nanowire structure, which has relatively smaller diameter compared to nanorods [69]. Other studies that applying a small potential range (0.3–0.6 V *vs.* SCE) also led to the formation of nanorod structure with similar diameter, that is around 30 nm in average [70]. This suggests the tendency to form nanorod structured manganese oxide at 0.3 V of potential range in electrolyte consisting manganese acetate and sodium sulfate. The specific capacitances estimated from charge–discharge at 0.1 mA cm^{-2} for nanorods formed in the potential ranges of (0.1–0.4) V and (0.3–0.6) V are in the range of 200 to 250 F g^{-1}. The extension of deposition potential range from 0.3 to 0.5 V not only improves the surface area, at the same time it also enhances the specific capacitance for 44%.

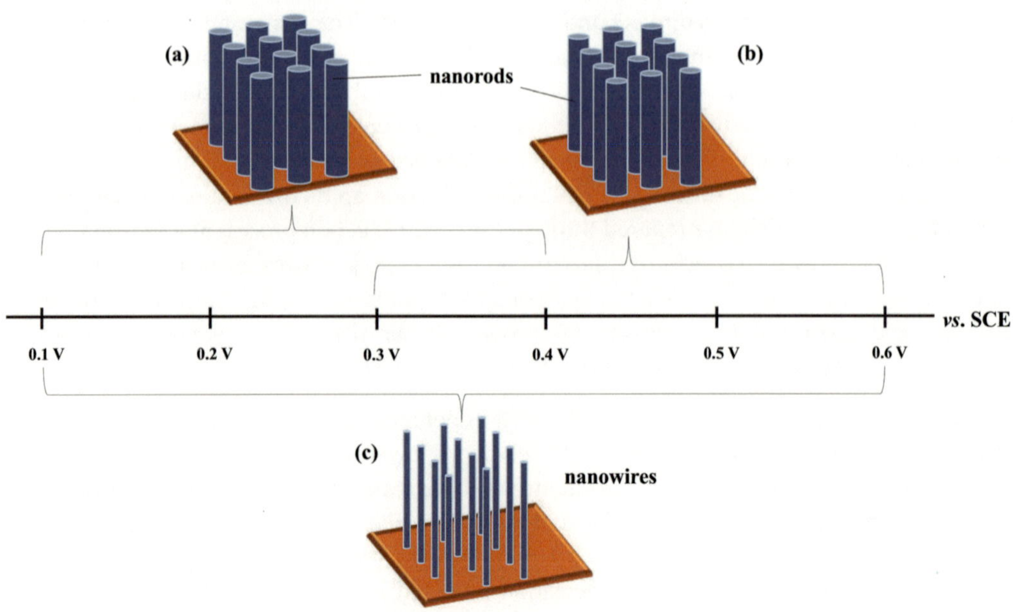

Figure 9. Morphologies formed at different potential range.

Although the integration with other materials shown enhancement in charge storage, the incorporation of carbon nanotubes (CNT) does not always work in this way. As the sp^2 carbon basal plane has low chemical and electrochemical reactivity, the nucleation sites on the surface of CNT-manganese oxide composite film are limited. In addition, the manganese oxide particles tend to grow at the CNT's junction [71]. A flower-like morphology built up by nanosheets that are originated radially from a central is thus formed. Nevertheless, this composite indeed acquires better cycle stability due to CNT acting as a conductive backbone that reduces the dissolution of particles. There is a variety of substrates can be used for deposition. Tantalum (Ta) foil has high melting point and provides good corrosion resistance, strength, and ductility. Cotton sheet supplies flexibility and textile structure that eases a uniform coating of material. Nickel (Ni) foil contributes a good electrical conductivity while the stainless steel is inert and owns a stable passivity. As a metal substrate, Ta foil and Ni foil usually offer a good cycle stability and specific capacitance of around 413 F g^{-1} in average for CNT-manganese oxide film [72, 73]. In spite of the fact that stainless steel has poorer electrical conductivity compared to other metals, the CNT-manganese oxide film composite film formed on stainless steel was found to achieve higher specific capacitance, that is 869 F g^{-1} [74]. Meanwhile, it also has good cycle stability. Given a similar morphology, which is nanowires-structured manganese oxide coated on CNT, formed on these three substrates, the distinct specific capacitance value achieved suggests the hidden advantages of stainless steel as a substrate. Direct deposition of manganese oxide on CNT paper using potentiodynamic method gives rise to around 168 F g^{-1} [70]. It is therefore important to choose an appropriate substrate for optimal electrochemical performance. Without the CNT, the manganese oxide tends to appear in nanorod-like structure and sphere-like structure at oxidation condition and reduction condition, respectively [75]. For the manganese oxide deposited within the same potential range (0.1−0.4 V), a desired structure can be determined by applying different annealing temperatures. For example, 300°C leads to the formation of nanotubes structure while 100°C forms nanorods structure [76]. The manganese oxide-based films prepared at different scan rates during potentiodynamic deposition possess distinct morphologies. At scan rate lower than 100 mV s^{-1}, the morphology forms are irregular and dense. It starts to evolve and becomes more porous when the scan rate of deposition is increased. A nanoflake structure can be formed at scan rate of 200 mV s^{-1} [77]. This structure is beneficial for charge storage purpose at which it has offered specific capacitance of 410 F g^{-1}. The compact and non-porous structure produced at relatively scan rate achieved around 150 F g^{-1}.

4.3. Effect of deposition current density

Potentiostatic deposition tends to form a more compact structure compared to the galvanostatic deposition. This is due to the consistent deposition rate during potentiostatic deposition. As the potential is maintained throughout the deposition process, the deposition rate is sustained. In contrast, the potential varies during galvanostatic deposition in order to sustain the current supply. The structure formed is thus less compact and higher in surface area [78]. The difference in compactness contributes to 50% increment in specific capacitance value and 15% higher capacitance retention upon 5000 cycles for galvanostatic deposited manganese oxide film. During the galvanostatic deposition, the structure changes from irregular to regular and uniform structure with the increased in current density [79, 80]. When the current density exceeds the optimal value to deposit Mn^{4+}, a clustered structure which consists of soluble Mn^{6+} and/or Mn^{7+} may forms. Other than oxidation states, the morphology also changes with current density. The evolution of morphology is easily observed using field emission scanning electron microscopy (FESEM). As the nucleation rate is directly related to the current density, a lower current density produces lower nucleation rate. As a result, there is not many nuclei formed on the substrate surface and a continuous deposit layer is hard to be constructed [81]. In addition, the deposit usually possesses rough surface. Higher current density can lead to higher nucleation and growth rate. The deposit accumulates on the structure formed ahead and filling up the space or crack, which leads to the formation of a uniform coating, **Figure 10**.

At optimal current density, various structures such as nanowires, nanoflakes, nanosheets, and nanorods can be formed [57, 82, 83]. For example, manganese oxide formed from $MnSO_4$ precursor at 4 mA cm^{-2} presents as agglomerated clusters. A small decreased in current density, 3.7 mA cm^{-2}, brings about a grain-like structure constructed by nanowires [84]. 2 mA cm^{-2} of galvanostatic deposition in $Mn(NO_3)_2$ produces a flower-like structure made up of nanowires [83]. This manganese oxide film is capable to maintain 84% of specific capacitance after 1000 cycles.

4.4. Electrophoretic deposition

As one kind of the electrodeposition techniques, the oxidation and dissolution of the metal substrate during the anodic electrophoretic deposition is a concern. Thus, cathodic electrophoretic deposition is more favorable compared to its counterpart. Different from the electrodeposition techniques mentioned above, the electrophoretic deposition does not form the oxide during or after the process. Instead, the oxide is fabricated beforehand. The desired nanostructure can be manipulated using various preparation methods such as hydrothermal, chemical reduction, chemical precipitation, spray pyrolysis, and wet-chemical processes. The nanostructured oxide powder formed is then suspended in a dispersant electrolyte for further action, which is the electrophoretic deposition in this case. The structure of manganese oxide particles will be retained even after deposition, as depicted in **Figure 11**.

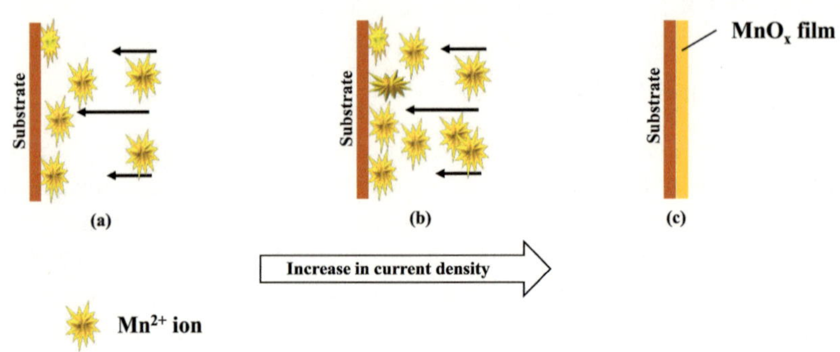

Figure 10. Morphological evolution with the increase in current density.

http://dx.doi.org/10.5772/intechopen.71957

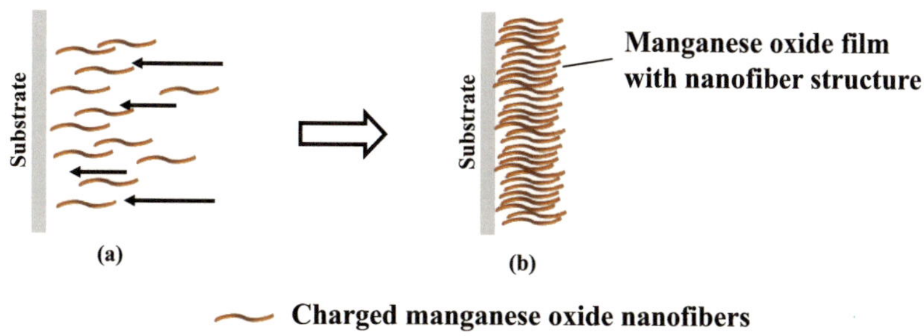

Figure 11. Schematic diagram of electrophoretic deposited manganese oxide film.

It can be seen that the stability of manganese oxide suspension is important to ensure the success of electrophoretic deposition. The ethanol can be used as the liquid medium for suspension. However, the deposit resulted from the manganese oxide suspension in ethanol is shown to be irregular and tends to form agglomerates. The addition of phosphate ester has been shown to enhance the stability of the manganese oxide suspension and increase the mass load [85]. This improves the electrochemical performance of the manganese oxide film. Without the phosphate ester, the manganese oxide film only exhibits 236 F g^{-1} of specific capacitance and drops 15% of the initial specific capacitance after 25 cycles [86]. Meanwhile, the phosphate ester has offered around 60% of increment in specific capacitance [85]. Sodium alginate is a good dispersant as well. It has been proposed that the sodium alginate provides the electrostatic and steric stabilization for the manganese oxide suspension. Additionally, it supplies electric charge for the suspension particles which is beneficial for the deposition process [87]. The manganese oxide film deposited from the dispersant electrolyte of sodium alginate has obtained specific capacitance (412 F g^{-1}) a little higher than the one prepared from the ethanol with phosphate ester dispersant electrolyte (377 F g^{-1}). The integration of carbon nanotubes or reduced graphene oxide with the manganese oxide does not alter the nanostructure of the oxide itself [88]. The nanostructured manganese oxide particles attach on the carbon nanotube surface and deposit together onto the substrate [89].

We have discussed about the impact of few deposition parameters on the deposited manganese oxide-based film. The as-deposited manganese oxide-based film is dominantly amorphous with inherent cation deficiency [90]. The defect is most likely to form at relatively low deposition temperatures (80–200°C). Except the one prepared using electrophoretic deposition, at which the manganese oxide particles are firstly fabricated before deposition, the as-deposited manganese oxide-based film requires certain post-treatments to improve the crystalline structure. Annealing is one of the common post-treatment. When the manganese oxide-based film undergoes post-heating process, different crystalline structure forms based on the annealing temperature. Most of the water content at the surface layer of structure desorbs gradually at 120–350°C [91]. The manganese oxide presents as γ-manganese oxide at annealing temperature lower than 350°C. The transformation of crystalline structure is initiated at around 300°C, which allows the γ-phase changes to β-phase of MnO_2. When the manganese oxide undergoes further heating, α-Mn_2O_3 phase starts to form [92]. Any desired crystalline structure is thus can be prepared. Other than crystalline structure, the morphology

Figure 12. Morphologies of manganese oxide film formed at annealing temperature of (a) 100 – 200°C, (b) 300°C, (c) 500°C, and (d) 600°C.

of the manganese oxide-based film can also be modified by varying the annealing temperatures. At annealing temperatures lower than 200°C, a fibrous and granular structure is formed. The entanglement of fibers causes the morphology to evolve to a cluster-like structure when the temperature reaches 300°C. Further increasing in temperature can lead to the formation of flaky-like (500°C) and rod-like structures (600°C) [93], **Figure 12**.

There are still many studies carried out to investigate other possible factors affecting the deposited manganese oxide-based film. For example, by studying the porosity of the manganese oxide-based film, one can also inspect more details about the structure and charge transport properties [94]. More and more deposition techniques have been developed and studied to prepare the manganese oxide-based film. For instance, a redox deposition that is took place when a substrate is immersed substrate into the Mn ions precursors [95, 96]. From chemical bath deposition, chemical vapor deposition to spray pyrolysis deposition, all deposition techniques are intended to grow the deposit with a good quality along with good physical and chemical properties in a large scale.

5. Summary

Manganese oxide-based film is shown to be prepared in various electrodeposition conditions by varying electrodeposition potentials, current densities, additives, and electrolytes. The potential application of manganese oxide-based film as the electrode material for electrochemical capacitor is thus discussed. As deposited manganese oxide film is amorphous in nature [97], the amorphousness can be transformed to crystalline phase by employing appropriate annealing temperature. The crystallinity start to arise when the annealing temperature increases to 300°C. MnO_2 is the first crystal structure detected at 300°C, further increasing the temperature leads to the formation of Mn_3O_4 and Mn_2O_3 [98]. Since proton participates in the charge storage mechanism of manganese oxide, it is believed that this oxide will perform better in hydrous form. Previous study about the RuO_2 has shown the significant improvement on the electrochemical performance with the presence of hydrous phase [99]. Not long after that, the crystalline manganese oxide has exhibited its potential application as electrode material as well [12]. To date, the charge storage mechanism is found to be dependent on the crystalline structure, water content, and surface area. It turns out that, the electrochemical performance of manganese oxide-based film is not totally relied on any of these factors. In contrast, it is resulted from the combination of all the factors that have been found and studied. For this reason, the researchers are still making effort to understand this complication.

The addition of secondary or ternary materials can enhance the electrochemical performance of manganese oxide film. However, these are not shown in this chapter. The core idea is to gather the advantages of various materials then compensate shortcomings of each other. The common materials combination involves the carbon material, transition metal oxide, and conducting polymer. The carbon material offers conductivity while transition metal oxide and conducting polymer provides more electroactive sites for charge storage purpose. Electrodeposition is a widely used technique to protect and strengthen the function of parts used in various industries. The wide application of electrodeposition technology can be attributed to its simplicity, manufacturability, and scalability. The electrodeposition method also requires a relatively low fabrication cost for energy storage device compared to other methods. This technique allows a direct formation of film on the substrate desired and the film properties are governable by varying the deposition parameters. The traditional method of electrode fabrication for electrochemical capacitor involves the pressing of electrode material onto the substrate. This process can increase the contact resistance and reduce the porous surface area which brings about damage to the electrode materials formed. In overall, manganese oxide-based film prepared using electrodeposition is prospective and practical. There are numerous related studies carrying out every year, although the reported specific capacitance of manganese oxide-based film are yet far from the expected performances (**Table 1**), we are convinced that there is still a big room for improvement.

Electrode materials	Deposition mode		Deposition electrolyte	Specific capacitance, F g^{-1}	Reference
MnO_x	Potentiostatic	0.2 V *vs.* SCE	0.001 M $KMnO_4$ + 1 M Na_2SO_4	368.00	[62]
MnO_x	Potentiostatic	0.50 V *vs.* SCE	0.25 M $Mn(CH_3COO)_2$	240.00	[63]
MnO_2	Potentiostatic	0.55 V *vs.* SCE	$KMnO_4$	232.94	[67]
a-$MnO_2{\cdot}nH_2O$	Potentiostatic	0.75 V *vs.* SCE	0.25 M $MnSO_4{\cdot}5H_2O$	285.00	[65]
MnO_x	Potentiostatic	1.0 V *vs.* Ag/AgCl	2 mM $MnSO_4$ + 50 mM KCl	163.40	[44]
MnO_x	Potentiostatic	1.1 V *vs.* Ag/AgCl	0.1 M $MnSO_4$ + 0.01 M TTAB	343.00	[100]
MnO_x			0.1 M $MnSO_4$	294.00	
MnO_2	Potentiostatic	10 V *vs.* Ag/AgCl	0.5 M $KMnO_4$	128.00	[78]
Mn_3O_4	Potentiostatic	−1.3 V *vs.* SCE	0.25 M $MnNO_3$	416.00	[101]
MnO_2	Potentiostatic	0.60 *vs.* SCE	0.1 M $Mn(CH_3COO)_2$ + 0.1 M Na_2SO_4	240.00	[102]
MnO_2-CNT	Cyclic voltammetry	0.30–0.60 V *vs.* SCE		167.50	[70]
Mn-Ni mixed oxide	Cyclic voltammetry	0.4 − 1.2 V *vs.* Ag/AgCl	0.05 M $Mn(CH_3COO)_2$ + 0.1 M $Ni(CH_3COO)_2$ + 0.2 M CH_3COONa	169.00	[103]
Mn-Mo mixed oxide	Cyclic voltammetry	0.00 − 1.00 V *vs.* Ag/AgCl	2 mM $MnSO_4$ + 20 mM Na_2MoO_4	190.90	[104]
MnO_x	Cyclic voltammetry	1.10 − 1.50 V *vs.* SCE	0.5 M $MnSO_4$ + 0.5 M Na_2SO_4 + 100 mM SLS	310.00	[105]

Electrode Materials	Deposition mode		Electrolyte composition	Specific capacitance, F g^{-1}	Reference
MnO_2	Galvanostatic	2 mA cm^{-2}	5 mM $Mn(NO_3)_2$	246.00	[83]
MnO_2	Galvanostatic	3 mA cm^{-2}	0.02 M $KMnO_4$	188.00	[106]
MnO_2-PPy	Galvanostatic	4 mA cm^{-2}	0.2 M $MnSO_4$ + PPy	620.00	[84]
Fe-doped MnO_x	Galvanostatic	5 mA cm^{-2}	0.1 M $MnSO_4$ + 0.1 M citric acid	218.00	[107]
MnO_x	Galvanostatic	5 mA cm^{-2}	0.01 M $Mn(CH_3COO)_2$	185.00	[57]
MnO_2	Galvanostatic	10.5 mA cm^{-2}	0.02 M $Mn(CH_3COO)_2$	201.00	[108]
Co-doped MnO_x	Galvanostatic	50 mA cm^{-2}	MnSO4, cobalt sulfate, EDTA	186.20	[109]
Fe-doped MnO_x	Galvanostatic		MnSO4, iron sulfate, EDTA	298.40	
MnO_2	Galvanostatic	165 mA cm^{-2}	0.5 M $KMnO_4$	196.00	[78]
MnO_2-RGO	Electrophoretic deposition	0.3 V *vs.* SCE	MnO_2 + RGO	392.00	[88]
MnO_X	Electrophoretic deposition	5 – 10 V *vs.* SCE	Manganese oxide + sodium alginate	412.00	[87]
MnO_x	Electrophoretic deposition	10 – 100 V *vs.* SCE	Manganese oxide + ethanol + phosphate ester	377.00	[85]
MnO_x-CNT	Electrophoretic deposition	15 V *vs.* SCE	Manganese oxide + sodium alginate + carbon nanotubes	Around 210.00	[89]
MnO_x	Electrophoretic deposition	100 V *vs.* Ag/AgCl	Manganese oxide + ethanol	236.00	[86]
MnO_x	Electrophoretic deposition	100 V *vs.* Ag/AgCl	Manganese oxide + ethylene alcohol + H_2SO_4	275.00	[110]

Table 1. The electrochemical performance of deposited manganese oxide-based film.

Acknowledgements

The authors acknowledge the University of Malaya for providing financial support through the projects BKS030-2017 and FG034-17AFR.

Author details

Chan Pei Yi and Siti Rohana Majid*

*Address all correspondence to: shana@um.edu.my

Centre for Ionics University of Malaya, Department of Physics, Faculty of Science, University of Malaya, Kuala Lumpur, Malaysia

References

[1] Ogden JM, Steinbugler MM, Kreutz TG. A comparison of hydrogen, methanol and gasoline as fuels for fuel cell vehicles: Implications for vehicle design and infrastructure development. Journal of Power Sources. 1999;**79**(2):143-168

[2] Lai JS, Nelson DJ. Energy management power converters in hybrid electric and fuel cell vehicles. Proceedings of the IEEE. 2007;**95**(4):766-777

[3] Garcia P, Fernandez LM, Garcia CA, Jurado F. Energy management system of fuel-cell-battery hybrid tramway. IEEE Transactions on Industrial Electronics. 2010;**57**(12):4013-4023

[4] Tsukahara K, Kondo K. A Study on Methods to Design and Select Energy Storage Devices for Fuel Cell Hybrid Powered Railway Vehicles. Paper presented at the IECON 2013-39th Annual Conference of the IEEE Industrial Electronics Society; 2013, 10-13 Nov. 2013

[5] Abraham KM. Prospects and limits of energy storage in batteries. Journal of Physical Chemistry Letters. 2015;**6**(5):830-844

[6] Braga MH, Grundish NS, Murchison AJ, Goodenough JB. Alternative strategy for a safe rechargeable battery. Energy & Environmental Science. 2017;**10**(1):331-336

[7] Ragone DV. Review of Battery Systems for Electrically Powered Vehicles; 1968

[8] McCloskey BD. Expanding the Ragone plot: Pushing the limits of energy storage. Journal of Physical Chemistry Letters. 2015;**6**(18):3592-3593

[9] Conway BE. Electrochemical Supercapacitors: Scientific Fundamentals and Technological Applications. US: Springer; 1999

[10] Zheng JP, Cygan PJ, Jow TR. Hydrous ruthenium oxide as an electrode material for electrochemical capacitors. Journal of the Electrochemical Society. 1995;**142**(8):2699-2703

[11] Brezesinski T, Wang J, Polleux J, Dunn B, Tolbert SH. Templated Nanocrystal-based porous TiO_2 films for next-generation electrochemical capacitors. Journal of the American Chemical Society. 2009;**131**(5):1802-1809

[12] Brousse T, Toupin M, Dugas R, Athouël L, Crosnier O, Bélanger D. Crystalline MnO_2 as possible alternatives to amorphous compounds in electrochemical supercapacitors. Journal of the Electrochemical Society. 2006;**153**(12):A2171-A2180

[13] Deng L, Zhang G, Kang L, Lei Z, Liu C, Liu Z-H. Graphene/VO_2 hybrid material for high performance electrochemical capacitor. Electrochimica Acta. 2013;**112**:448-457

[14] Li F, Song J, Yang H, Gan S, Zhang Q, Han D, et al. One-step synthesis of graphene/SnO_2 nanocomposites and its application in electrochemical supercapacitors. Nanotechnology. 2009;**20**(45):455602

[15] Xiao X, Ding T, Yuan L, Shen Y, Zhong Q, Zhang X, et al. WO3−x/MoO3−x Core/Shell nanowires on carbon fabric as an anode for all-solid-state asymmetric supercapacitors. Advanced Energy Materials. 2012;**2**(11):1328-1332

[16] Sin D-Y, Koo B-R, Ahn H-J. Hollow lithium manganese oxide nanotubes using MnO_2-carbon nanofiber composites as cathode materials for hybrid capacitors. Journal of Alloys and Compounds. 2017;**696**:290-294

[17] Unnikrishnan B, Wu C-W, Chen IWP, Chang H-T, Lin C-H, Huang C-C. Carbon dot-mediated synthesis of manganese oxide decorated graphene nanosheets for supercapacitor application. ACS Sustainable Chemistry & Engineering. 2016;**4**(6): 3008-3016

[18] Wang Y, Zhitomirsky I. Cathodic electrodeposition of Ag-doped manganese dioxide films for electrodes of electrochemical supercapacitors. Materials Letters. 2011;**65** (12):1759-1761

[19] Maynard JB. The chemistry of manganese ores through time: A signal of increasing diversity of earth-surface environments. Economic Geography. 2010;**105**(3):535-552

[20] Augustin M, Fenske D, Bardenhagen I, Westphal A, Knipper M, Plaggenborg T, et al. Manganese oxide phases and morphologies: A study on calcination temperature and atmospheric dependence. Beilstein Journal of Nanotechnology. 2015;**6**:47-59

[21] Feng Q, Kanoh H, Ooi K. Manganese oxide porous crystals. Journal of Materials Chemistry. 1999;**9**(2):319-333

[22] Jia M-Y, Xu B, Ding X-L, He S-G, Ge M-F. Experimental and theoretical study of the reactions between manganese oxide cluster anions and hydrogen sulfide. Journal of Physical Chemistry C. 2012;**116**(45):24184-24192

[23] Kulp JL, Perfetti JN. Thermal study of some manganese oxide minerals. Mineralogical Magazine. 1950;**29**(210):239-252

[24] Mande C, Deshpande AP. A study of manganese oxides by EXAFS spectroscopy. Physica Status Solidi (b). 1990;**158**(2):737-742

[25] Post JE. Manganese oxide minerals: Crystal structures and economic and environmental significance. Proceedings of the National Academy of Sciences of the United States of America. 1999;**96**(7):3447-345 4

[26] Tian Z-R, Tong W, Wang J-Y, Duan N-G, Krishnan VV, Suib SL. Manganese oxide mesoporous structures: Mixed-valent semiconducting catalysts. Sci. 1997;**276**(5314):926-930

[27] Devaraj S, Munichandraiah N. Effect of crystallographic structure of MnO_2 on its electrochemical capacitance properties. Journal of Physical Chemistry C. 2008;**112**(11):4406-4417

[28] Bigliocca C, Girardi F, Pauly J, Sabbioni E, Meloni S, Provasoli A. Radiochemical separations by adsorption on manganese dioxide. Analytical Chemistry. 1967;**39**(13):1634-1639

[29] Li L, Xu J, Zheng X, Ma C, Song X, Ge S, et al. Growth of gold-manganese oxide nanostructures on a 3D origami device for glucose-oxidase label based electrochemical immunosensor. Biosensors & Bioelectronics. 2014;**61**:76-82

[30] Li X, Hu B, Suib S, Lei Y, Li B. Manganese dioxide as a new cathode catalyst in microbial fuel cells. Journal of Power Sources. 2010;**195**(9):2586-2591

[31] Loganathan P, Burau RG. Sorption of heavy metal ions by a hydrous manganese oxide. Geochimica et Cosmochimica Acta. 1973;**37**(5):1277-1293

[32] Piispanen JK, Sallanko JT. Mn(II) removal from groundwater with manganese oxide-coated filter media. Journal of Environmental Science and Health, Part A. 2010;**45**(13): 1732-1740

[33] Smith PF, Deibert BJ, Kaushik S, Gardner G, Hwang S, Wang H, et al. Coordination geometry and oxidation state requirements of corner-sharing MnO_6 octahedra for water oxidation catalysis: An investigation of manganite (γ-MnOOH). ACS Catalysis. 2016;**6**(3): 2089-2099

[34] Thackeray MM. Manganese oxides for lithium batteries. Progress in Solid State Chemistry. 1997;**25**(1):1-71

[35] Pourbaix M. Atlas of Electrochemical Equilibria in Aqueous Solutions Retrieved from ElectronicBooks database; 1974

[36] Pang SC, Anderson MA, Chapman TW. Novel electrode materials for thin film ultracapacitors: Comparison of electrochemical properties of sol gel derived and electrodeposited manganese dioxide. Journal of the Electrochemical Society. 2000;**147**(2):444-450

[37] Dupont M, Hollenkamp AF, Donne SW. Electrochemically active surface area effects on the performance of manganese dioxide for electrochemical capacitor applications. Electrochimica Acta. 2013;**104**:140-147

[38] Ghaemi M, Ataherian F, Zolfaghari A, Jafari SM. Charge storage mechanism of sonochemically prepared MnO_2 as supercapacitor electrode: Effects of physisorbed water and proton conduction. Electrochimica Acta. 2008;**53**(14):4607-4614

[39] Toupin M, Brousse T, Bélanger D. Influence of microstucture on the charge storage properties of chemically synthesized manganese dioxide. Chemistry of Materials. 2002;**14**(9): 3946-3952

[40] Wu T-H, Hesp D, Dhanak V, Collins C, Braga F, Hardwick LJ, Hu C-C. Charge storage mechanism of activated manganese oxide composites for pseudocapacitors. Journal of Materials Chemistry A. 2015;**3**(24):12786-12795

[41] Ahmed KAM, Peng H, Wu K, Huang K. Hydrothermal preparation of nanostructured manganese oxides (MnOx) and their electrochemical and photocatalytic properties. Chemical Engineering Journal. 2011;**172**(1):531-539

[42] Aref AA, Tang YW. Chemical bath deposition synthesis and electrochemical properties of MnO_2 thin film: Effect of deposition time and bath temperature. Materials Science-Poland. 2014;**32**(4):555-564

[43] Lee SW, Kim J, Chen S, Hammond PT, Shao-Horn Y. Carbon nanotube/manganese oxide ultrathin film electrodes for electrochemical capacitors. ACS Nano. 2010;**4**(7):3889-3896

[44] Nakayama M, Kanaya T, Inoue R. Anodic deposition of layered manganese oxide into a colloidal crystal template for electrochemical supercapacitor. Electrochemistry Communications. 2007;**9**(5):1154-1158

[45] Ni J, Lu W, Zhang L, Yue B, Shang X, Lv Y. Low-temperature synthesis of monodisperse 3D manganese oxide nanoflowers and their pseudocapacitance properties. Journal of Physical Chemistry C. 2009;**113**(1):54-60

[46] Stumm W, Morgan JJ. Aquatic chemistry: an introduction emphasizing chemical equilibria in natural waters (pp. 795-795); 1981

[47] Wang Y, Zhitomirsky I. Electrophoretic deposition of manganese dioxide–multiwalled carbon nanotube composites for electrochemical supercapacitors. Langmuir. 2009;**25** (17):9684-9689

[48] Xiao XF, Liu RF. Effect of suspension stability on electrophoretic deposition of hydroxyapatite coatings. Materials Letters. 2006;**60**(21–22):2627-2632

[49] Besra L, Liu M. A review on fundamentals and applications of electrophoretic deposition (EPD). Progress in Materials Science. 2007;**52**(1):1-61

[50] Zhitomirsky I. Cathodic electrodeposition of ceramic and organoceramic materials. Fundamental aspects. Advances in Colloid and Interface Science. 2002;**97**(1–3):279-317

[51] Zhitomirsky I, Cheong M, Wei J. The cathodic electrodeposition of manganese oxide films for electrochemical supercapacitors. JOM. 2007;**59**(7):66-69

[52] Chen Y-S, Hu C-C, Wu Y-T. Capacitive and textural characteristics of manganese oxide prepared by anodic deposition: Effects of manganese precursors and oxide thickness. Journal of Solid State Electrochemistry. 2004;**8**(7):467-473

[53] Rangappa KS, Chandraju D, Made Gowda NM. Oxidation of L-glutamine by manganese (III) in aqueous sulfuric acid, acetic acid, and pyrophosphate media: A kinetic and mechanistic study. International Journal of Chemical Kinetics. 1998;**30**:7-19

[54] Guidelli R, Piccardi G. The voltammetric behaviour of the Mn2+, Mn3+, Mn4+ system in 15 N H2SO4 on a smooth platinum microelectrode. Electrochimica Acta. 1968;**13**(1): 99-107

[55] Kao W-H, Weibel VJ. Electrochemical oxidation of manganese(II) at a platinum electrode. Journal of Applied Electrochemistry. 1992;**22**(1):21-27

[56] Fleischmann M, Thirsk HR, Tordesillas IM. Kinetics of electrodeposition of γ-manganese dioxide. Transactions of the Faraday Society. 1962;**58**(0):1865-1877

[57] Babakhani B, Ivey DG. Anodic deposition of manganese oxide electrodes with rod-like structures for application as electrochemical capacitors. Journal of Power Sources. 2010;**195**(7):2110-2117

[58] Dupont MF, Donne SW. Nucleation and growth of electrodeposited manganese dioxide for electrochemical capacitors. Electrochimica Acta. 2014;**120**:219-225

[59] Cross A, Morel A, Cormie A, Hollenkamp T, Donne S. Enhanced manganese dioxide supercapacitor electrodes produced by electrodeposition. Journal of Power Sources. 2011;**196**(18):7847-7853

[60] Clarke CJ, Browning GJ, Donne SW. An RDE and RRDE study into the electrodeposition of manganese dioxide. Electrochimica Acta. 2006;**51**(26):5773-5784

[61] Broughton JN, Brett MJ. Variations in MnO_2 electrodeposition for electrochemical capacitors. Electrochimica Acta. 2005;**50**(24):4814-4819

[62] Gibson AJ, Latham KG, Burns RC, Donne SW. Electrodeposition mechanism of cathodically-prepared manganese dioxide thin films from permanganate for use in electrochemical capacitors. Electrochimica Acta. 2017;**236**:198-211

[63] Chang J-K, Tsai W-T. Material characterization and electrochemical performance of hydrous manganese oxide electrodes for use in electrochemical pseudocapacitors. Journal of the Electrochemical Society. 2003;**150**(10):A1333-A1338

[64] Chigane M, Ishikawa M. Manganese oxide thin film preparation by potentiostatic electrolyses and electrochromism. Journal of the Electrochemical Society. 2000;**147**(6): 2246-2251

[65] Hu C-C, Tsou T-W. Ideal capacitive behavior of hydrous manganese oxide prepared by anodic deposition. Electrochemistry Communications. 2002;**4**(2):105-109

[66] Rodrigues S, Shukla AK, Munichandraiah N. A cyclic voltammetric study of the kinetics and mechanism of electrodeposition of manganese dioxide. Journal of Applied Electrochemistry. 1998;**28**(11):1235-1241

[67] Yuqiu H, Hongcheng Z. Cathodic potentiostatic electrodeposition and capacitance characterization of manganese dioxide film. Energy Procedia. 2011;**11**:2769-2774

[68] Liu D, Garcia BB, Zhang Q, Guo Q, Zhang Y, Sepehri S, Cao G. Mesoporous hydrous manganese dioxide nanowall arrays with large lithium ion energy storage capacities. Advanced Functional Materials. 2009;**19**(7):1015-1023

[69] Wu M-S. Electrochemical capacitance from manganese oxide nanowire structure synthesized by cyclic voltammetric electrodeposition. Applied Physics Letters. 2005;**87**(15): 153102

[70] Chou S-L, Wang J-Z, Chew S-Y, Liu H-K, Dou S-X. Electrodeposition of MnO_2 nanowires on carbon nanotube paper as free-standing, flexible electrode for supercapacitors. Electrochemistry Communications. 2008;**10**(11):1724-1727

[71] Zhang H, Cao G, Wang Z, Yang Y, Shi Z, Gu Z. Growth of manganese oxide nanoflowers on vertically-aligned carbon nanotube arrays for high-rate electrochemical capacitive energy storage. Nano Letters. 2008;**8**(9):2664-2668

[72] Hu L, Chen W, Xie X, Liu N, Yang Y, Wu H, et al. Symmetrical MnO_2–carbon nanotube–textile nanostructures for wearable pseudocapacitors with high mass loading. ACS Nano. 2011;**5**(11):8904-8913

[73] Lee CY, Tsai HM, Chuang HJ, Li SY, Lin P, Tseng TY. Characteristics and electrochemical performance of Supercapacitors with manganese oxide-carbon nanotube nanocomposite electrodes. Journal of the Electrochemical Society. 2005;**152**(4):A716-A720

[74] Bordjiba T, Bélanger D. Development of new nanocomposite based on nanosized-manganese oxide and carbon nanotubes for high performance electrochemical capacitors. Electrochimica Acta. 2010;**55**(9):3428-3433

[75] Ashassi-Sorkhabi H, Asghari E, La'le Badakhshan P. Potentiostatic and cyclic voltammetric deposition of nanostructured manganese oxide for supercapacitor applications. Current Applied Physics. 2014;**14**(2):187-191

[76] Wu M-S, Lee J-T, Wang Y-Y, Wan C-C. Field emission from manganese oxide nanotubes synthesized by cyclic voltammetric electrodeposition. The Journal of Physical Chemistry. B. 2004a;**108**(42):16331-16333

[77] Prasad KR, Miura N. Potentiodynamically deposited nanostructured manganese dioxide as electrode material for electrochemical redox supercapacitors. Journal of Power Sources. 2004;**135**(1–2):354-360

[78] Ali GAM, Yusoff MM, Ng YH, Lim HN, Chong KF. Potentiostatic and galvanostatic electrodeposition of manganese oxide for supercapacitor application: A comparison study. Current Applied Physics. 2015;**15**(10):1143-1147

[79] Drosos H, Sapountzis A, Koudoumas E, Katsarakis N, Vernardou D. Effect of deposition current density on electrodeposited vanadium oxide coatings. Journal of the Electrochemical Society. 2012;**159**(8):E145-E147

[80] Guo HJ, Zhu BQ, Li XH, Zhang XM, Wang ZX, Peng WJ, Liu LP. Effects of current density on preparation of grainy electrolytic manganese dioxide. Journal of Central South University. 2005;**12**(6):667-670

[81] Babakhani B, Ivey DG. Effect of electrodeposition conditions on the electrochemical capacitive behavior of synthesized manganese oxide electrodes. Journal of Power Sources. 2011;**196**(24):10762-10774

[82] Wei W, Cui X, Mao X, Chen W, Ivey DG. Morphology evolution in anodically electrodeposited manganese oxide nanostructures for electrochemical supercapacitor applications—Effect of supersaturation ratio. Electrochimica Acta. 2011;**56**(3):1619-1628

[83] Yousefi T, Golikand AN, Mashhadizadeh MH, Aghazadeh M. Template-free synthesis of MnO_2 nanowires with secondary flower like structure: Characterization and supercapacitor behavior studies. Current Applied Physics. 2012;**12**(1):193-198

[84] Sharma RK, Rastogi AC, Desu SB. Manganese oxide embedded polypyrrole nanocomposites for electrochemical supercapacitor. Electrochimica Acta. 2008;**53**(26):7690-7695

[85] Li J, Zhitomirsky I. Cathodic electrophoretic deposition of manganese dioxide films. Colloids and Surfaces A: Physicochemical and Engineering. 2009a;**348**(1):248-253

[86] Chen C-Y, Wang S-C, Lin C-Y, Chen F-S, Lin C-K. Electrophoretically deposited manganese oxide coatings for supercapacitor application. Ceramics International. 2009;**35**(8): 3469-3474

[87] Li J, Zhitomirsky I. Electrophoretic deposition of manganese oxide nanofibers. Materials Chemistry and Physics. 2008;**112**(2):525-530

[88] Ghasemi S, Hosseinzadeh R, Jafari M. MnO_2 nanoparticles decorated on electrophoretically deposited graphene nanosheets for high performance supercapacitor. International Journal of Hydrogen Energy. 2015;**40**(2):1037-1046

[89] Li J, Zhitomirsky I. Electrophoretic deposition of manganese dioxide–carbon nanotube composites. The Journal of Materials Processing Technology. 2009b;**209**(7):3452-3459

[90] Wei W, Chen W, Ivey DG. Defective rock-salt structure in anodically electrodeposited Mn−co−O nanocrystals. Journal of Physical Chemistry C. 2007;**111**(28):10398-10403

[91] Preisler E. Semiconductor properties of manganese dioxide. Journal of Applied Electrochemistry. 1976;**6**(4):311-320

[92] Giovanoli R, Stähli E, Feitknecht W. Über oxidhydroxide des vierwertigen Mangans mit Schichtengitter 2. Mitteilung: Mangan (III)-manganat (IV). Helvetica Chimica Acta. 1970;**53**(3):453-464

[93] Chang J-K, Chen Y-L, Tsai W-T. Effect of heat treatment on material characteristics and pseudo-capacitive properties of manganese oxide prepared by anodic deposition. Journal of Power Sources. 2004;**135**(1–2):344-353

[94] Gamble T, Gillette E, Lee SB, Siwy ZS. Probing porous structure of single manganese oxide mesorods with ionic current. Journal of Physical Chemistry C. 2013;**117**(47):24836-24842

[95] George AT, Ganesan R, Thangeeswari T. Redox Deposition of Manganese Oxide Nanoparticles on Graphite Electrode by Immersion Technique for Electrochemical Super Capacitors; 2016

[96] Wu M, Snook GA, Chen GZ, Fray DJ. Redox deposition of manganese oxide on graphite for supercapacitors. Electrochemistry Communications. 2004b;**6**(5):499-504

[97] Shinomiya T, Gupta V, Miura N. Effects of electrochemical-deposition method and microstructure on the capacitive characteristics of nano-sized manganese oxide. Electrochimica Acta. 2006;**51**(21):4412-4419

[98] Nagarajan N, Humadi H, Zhitomirsky I. Cathodic electrodeposition of MnOx films for electrochemical supercapacitors. Electrochimica Acta. 2006;**51**(15):3039-3045

[99] Zheng J, Huang C. Electrochemical behavior of amorphous and crystalline ruthenium oxide electrodes. Journal of New Materials for Electrochemical Systems. 2002;**5**:41

[100] Lee H-M, Cho S-W, Song C-J, Kang HJ, Kwon BJ, Kim C-K. Abrupt change with surfactant concentration in the surface morphology of the electrodeposited manganese oxide films for electrochemical capacitors. Electrochimica Acta. 2015;**160**:50-56

[101] Nguyen T, João Carmezim M, Boudard M, Fátima Montemor M. Cathodic electrodeposition and electrochemical response of manganese oxide pseudocapacitor electrodes. International Journal of Hydrogen Energy. 2015;**40**(46):16355-16364

[102] Chou S, Cheng F, Chen J. Electrodeposition synthesis and electrochemical properties of nanostructured γ-MnO_2 films. Journal of Power Sources. 2006;**162**(1):727-734

[103] Tahmasebi MH, Raeissi K, Golozar MA, Vicenzo A, Hashempour M, Bestetti M. Tailoring the pseudocapacitive behavior of electrochemically deposited manganese-nickel oxide films. Electrochimica Acta. 2016;**190**:636-647

[104] Masaharu Nakayama AT, Sato Y, Tonosaki T, Ogura K. Electrodeposition of manganese and molybdenum mixed oxide thin films and their charge storage properties. Langmuir. 2005;**21**(13):5907-5913

[105] Devaraj S, Munichandraiah N. High capacitance of electrodeposited MnO_2 by the effect of a surface-active agent. Electrochemical and Solid-State Letters. 2005;**8**(7):A373-A377

[106] Jacob GM, Zhitomirsky I. Microstructure and properties of manganese dioxide films prepared by electrodeposition. Applied Surface Science. 2008;**254**(20):6671-6676

[107] Dubal DP, Kim WB, Lokhande CD. Galvanostatically deposited Fe: MnO_2 electrodes for supercapacitor application. Journal of Physics and Chemistry of Solids. 2012;**73**(1):18-24

[108] Kundu M, Liu L. Direct growth of mesoporous MnO_2 nanosheet arrays on nickel foam current collectors for high-performance pseudocapacitors. Journal of Power Sources. 2013;**243**:676-681

[109] Trung Dung D, Thi Thu Hang L, Thi Bich Thuy H, Thanh Tung M. Synthesis of nanostructured manganese oxides based materials and application for supercapacitor. Advances in Natural Sciences: Nanoscience and Nanotechnology. 2015;**6**(2):025011

[110] Chen C-Y, Lyu Y-R, Su C-Y, Lin H-M, Lin C-K. Characterization of spray pyrolyzed manganese oxide powders deposited by electrophoretic deposition technique. Surface and Coating Technology. 2007;**202**(4):1277-1281

Chapter 8

Semiconducting Electrospun Nanofibers for Energy Conversion

Giulia Massaglia and Marzia Quaglio

Additional information is available at the end of the chapter

http://dx.doi.org/10.5772/intechopen.72817

Abstract

Nowadays, semiconducting thin films, thanks to their unique and excellent properties, play a crucial role for the design of devices for energy conversion and storage, such as solar cells, perovskite solar cells, lithium-ion batteries (LIBs), and fuel cells. Since the nanostructured arrangements can improve the behavior of the materials in several application fields, in this chapter we propose the electrospinning process as electrohydrodynamic deposition to obtain semiconducting materials, in the form of nanofiber mats. The nanostructured mats are able to provide high surface-area-to-volume ratio and a microporous structure, which are crucial aspects for energetic application. In this chapter, we deeply describe the electrospinning process and how nanofibers obtained can be used in energy devices, satisfying all the requirements to improve overall final performances.

Keywords: semiconducting materials, nanofiber mats, electrospinning, energy conversion devices, energy storage devices

1. Introduction

Nowadays, different deposition techniques, based on the application of an external electric potential, are carried out in order to prepare thin film and coatings. A thin film is defined as a layer of material with a thickness in the range from few nanometers (namely monolayer) to several micrometers. All electrodeposition processes, which can be divided in chemical methods and physical methods as proposed in **Figure 1**, ensure the deposition of different classes of materials as metals, semiconductors, ceramics, and organo-ceramics in the form of thin films, onto several substrate materials. Semiconducting thin films show a variety of unique and excellent properties that make them particularly attractive in several application areas. Among them, these materials play a preeminent role for the design of devices for energy conversion, such as solar cells and perovskite solar cells [1]. The behavior of semiconducting materials can be improved toward energy-related applications, when their shape and dimension are controlled

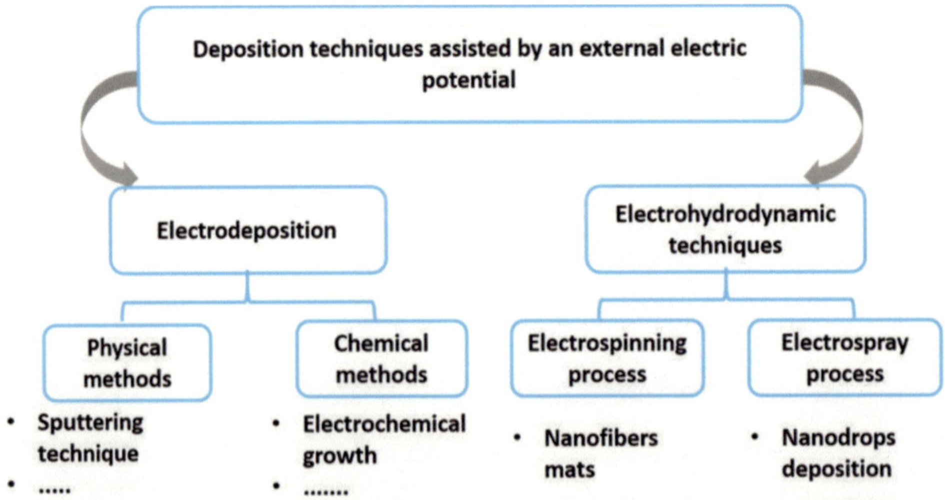

Figure 1. Different deposition techniques obtained by applying an external electric field and/or potential difference. These techniques can be divided in electrodeposition methods and electro-hydrodynamic techniques, as explained in the diagram.

up to the nanometer scale, in the so-called nanostructured arrangements. In this chapter, electro-hydrodynamic depositions are proposed in order to obtain semiconducting materials in the form of nanofiber mats, able to combine high surface-area-to-volume ratio together with a microporous structure well suited for energetic application. The electro-hydrodynamic techniques involve an external electric field applied to a polymeric solution to provide the final deposition of nanomaterial. As proposed in **Figure 1**, two different processes, i.e., electrospinning and electrospray, can be classified as electro-hydrodynamic techniques. Indeed, electrospinning ensures the direct assembly of nanofiber mats with different morphologies and properties, as described in this chapter.

Electrospinning process is based on the principle that strong repulsive forces, induced by external applied electric field, can overcome the surface tension in a charged polymeric jet [2, 3]. Therefore, through this technique, the polymer can be arranged in a mat with a high surface-area-ratio-to-volume, showing a micro−/macroporous structure. Moreover, final nanofibers are based not only on polymers but also on metals, ceramics, and metal oxides, obtained by implementing further different chemical and thermal treatments. However, electrospray technique is an electro-hydrodynamic technique, which occurs at low viscosity values of initial polymeric solution. Indeed, in this case, the surface tension overcomes the viscoelastic forces, and consequently, the instauration of charged droplets with different diameters and concentrations occurs during the process.

During the last decade, different works have been presented in the literature, focusing their attention on nanostructured semiconducting metal oxides (as TiO_2, ZnO, CuO, and SnO_2) in order to design well-performing and green energy systems (such as in dye-sensitized solar cells, lithium-ion batteries (LIBs), fuel cells). In this scenario, nanofibers progressively increased their importance as one of the most important nanostructures to be selected to improve the final performance of the devices.

http://dx.doi.org/10.5772/intechopen.72817

2. Electrospinning technique and its principles

The electrospinning is an electro-hydrodynamic process that provides polymer-based fibers with diameter distribution in the range from few nanometers to several micrometers by involving electrostatic forces [4–7]. The process is based on the concept that electrostatic forces induce columbic interactions between charged elements of the polymeric fluid, leading then to overcome the surface tension in a charged polymeric jet and ensuring the nanofiber formation. An electrospinning system is constituted by three major components, as sketched in **Figure 2(a)**: (i) high-voltage supply; (ii) a spinneret, which represents one of the two electrodes, containing the metallic needle of the syringe, where the polymeric solution is loaded; and (iii) the counter electrode, also named grounded electrode, which is the second electrode, where the nanofibers are collected. It involves a high-voltage supply in order to inject charges with a certain polarity in the polymeric solution and then generates a polymeric charged jet, accelerated toward a counter electrode with opposite polarity. In a typical process, the voltage (0–30 kV) is applied between the first electrode (tip of needle) and the second electrode (counter electrode). This implies the indirectly definition of electric field intensity as the ratio between the voltage value and working distance. The working distance is the distance between the first electrode and the counter electrode. The spinneret is linked with a syringe, in which the polymeric (or melt) solution is loaded and a syringe pump allows to control the solution flows with a constant rate, defined as flow rate. When the voltage is applied, the drop at the tip of the needle becomes highly electrified, and the charges are uniformly distributed on its surface. Therefore, the repulsive forces, acted between all charged elements of polymeric solution, induce an elongation of the spherical drop to form a conical shape, known as Taylor's cone. When the repulsive forces

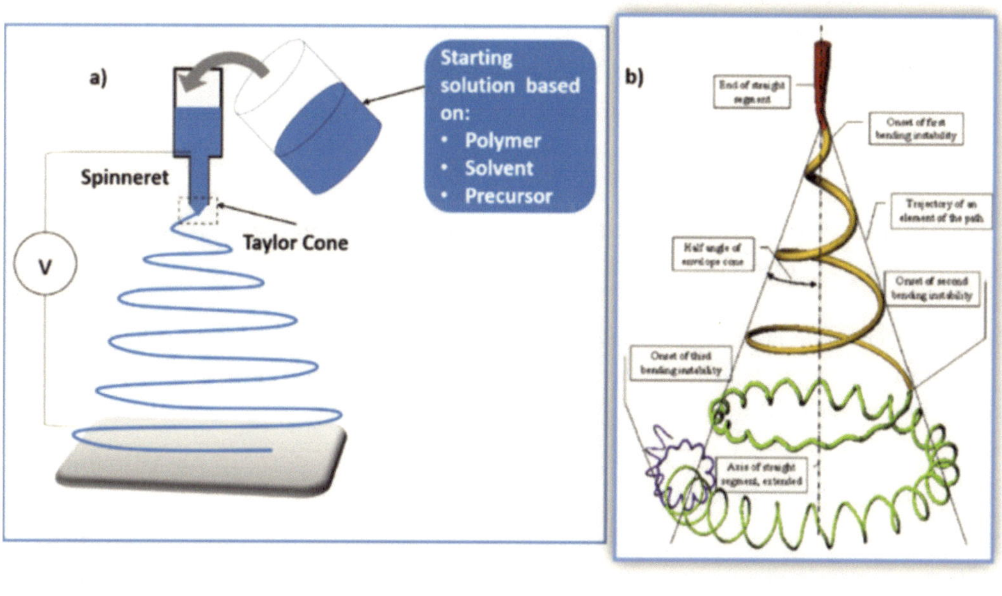

Figure 2. A sketch of electrospinning setup is proposed in (a). In (b) a representation of bending instabilities characterizing the charged polymeric jet during the electrospinning process is proposed. (reprinted with the permission from (polymer, 2008, 49, 2387–2425) copyright (2008) Elsevier).

overcame the surface tension of the droplet, the charged polymeric jet is ejected from the tip of Taylor's cone. During the flight, the solvent evaporation together with the instauration of several instabilities (defined whipping or bending instabilities) [8–10] occurred, leading then to the deposition of nanofiber mat, characterized by a small-size diameter distribution and by a high surface-area-to-volume ratio. In particular, the electrified jet proceeded with a straight path directly toward the counter electrode until the formation of successive instabilities, as sketched in **Figure 2(b)**.

The theoretical principle that explains the correlation between the formation of bending instabilities with the columbic interactions, the external electric field, and the surface tension is not widely investigated in the literature. However, during the process, the free end of the jet shows different envelope loops, which repeats itself in a smaller and smaller scale as the jet diameter is reduced [10]. During the bending instability, the charged jet is divided in sub jets, achieving a progressive diameter reduction, determined by Eq. (1) as explained in the literature [2]:

$$r_0{}^3 = \frac{4\varepsilon \dot{m}_0}{k\pi\sigma\rho} \tag{1}$$

where ε is the fluidic permittivity (C V^{-1} cm^{-1}), $\dot{m}_0$ is the mass flow rate (g s^{-1}) when r_0 (cm) is defined, k is a dimensionless parameter depending on the electric currents, σ is electric conductivity (A V^{-1} cm^{-1}), and ρ is the density (g cm^{-3}) of obtained nanofibers mats. Electrospinning process is applied on the polymer-based materials, including both synthetic and natural polymers. However, metallic carbon nanofibers and ceramic nanofibers can be obtained by electrospinning process, starting from polymeric solutions, and by occurring successive treatments, such as pyrolysis, calcination, and so on. One of the main advantages of this process is represented by the different nanostructures that can be obtained, such as hollow, porous, and dense nanofibers. Therefore, all wide nanostructures are achieved by varying and defining the process parameters, such as electric potential, flow rate, polymer concentration, working distance, and ambient condition.

3. Definition of electrospinning parameters and their correlation with the nanofiber properties

Since the modulation of morphological properties of the nanofiber mats is directly dependent on the process parameters, it is mandatory to define all these process parameters, which can be divided in three main categories [3, 11, 12]:

i. Parameters of the polymeric solution (or polymer melt), i.e., viscosity, concentration, and polymer molecular weight

ii. Parameters of electrospinning process, i.e., voltage, flow rate, and working distance between two electrodes

iii. External parameters, i.e., humidity and temperature

The first two categories are analyzed in the following paragraphs.

http://dx.doi.org/10.5772/intechopen.72817

3.1. Polymer solution parameters

Several solution parameters, such as viscosity solution, conductivity, dielectric constant, and surface tension, influence the formation of polymeric charged jet and consequently morphological properties of nanofiber mats. The solution viscosity represents the resistance offered by a fluid to its progressive deformation, induced by shear stress or tensile stress. In particular, the viscosity is due to the collisions between all particles that move in a fluid at different velocities. Therefore, solution viscosity can be defined as the measure of force/stress needed to keep the fluid moving in a certain space. The concentration of polymer, dissolved in the solution, directly influences its viscosity the higher the polymeric concentration, the higher the solution viscosity. As determined by several works in the literature, in order to guarantee the instauration of charged polymeric jet during the process, leading to the collection of nanofibers, the viscosity must be in the following range (Eq. (2)) [13–15]:

$$(0.02 \leq \eta \leq 300)\ Pa{*}s \tag{2}$$

The formation of nanofibers with or without defects depends on both viscosity and surface tension of the solution. The surface tension is a polymeric solution property due to all cohesive forces between fluidic molecules, ensuring then/leading then to the distribution of a fluid into the minimum surface area condition. Indeed, in one liquid all inner molecules interact with each neighboring molecule, inducing a resulting force equal to zero and a lower state of energy. On the contrary, since the same number of neighboring does not surround the molecules on the surface, an internal pressure is occurred, which induces the liquid surface to occupy the minimal area, reducing its energy state. According to Laplace's law, the spherical shape can satisfy conditions of minimal area for a liquid [3], minimizing the "wall tension" of the drop surface, as sketched in **Figure 3**. Related to the electrospinning process, the surface tension of polymeric solution ensures the generation of a spherical droplet, suspended at the tip of the needle.

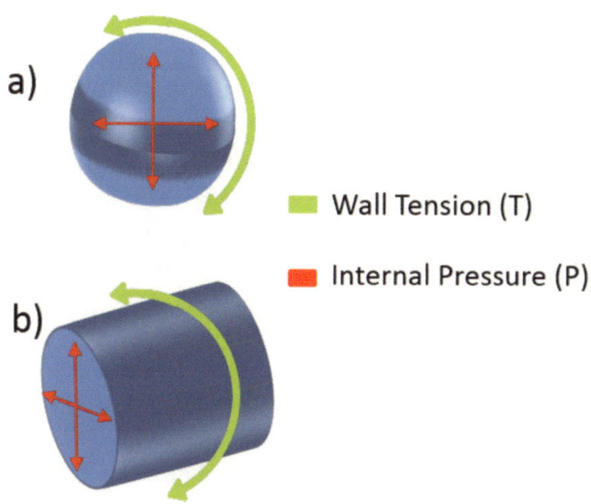

Figure 3. Representation of theoretical concept of Laplace's law (a) represents the cylindrical vessel (T = PR, where R is the radius of tube) and (b) represents the spherical vessel (T = PR/2, where R is the radius of tube).

In order to explain the correlation between the solution concentration, solution viscosity, and surface tension, it is important to distinguish two different electro-hydrodynamic processes, obtained by using polymeric solutions with different viscosity values. Indeed, if viscosity is lower than 0.1 Pa*s ($\eta \leq 0.1$ Pa*s), the surface tension overcomes the viscoelastic forces, a noncontinuous charged polymeric jet is generated, and consequently, droplets with different diameters and with different concentrations are collected. This process is defined as electrospray [13–15]. When viscosity value is higher than 2 Pa*s ($\eta \geq 2$ Pa*s), the electrospinning process is ensured, thus providing the formation of nanofibers. Therefore, the charged polymeric jet travels as a continuous jet toward counter electrode, in which dried nanofiber mats were collected on.

Different works in the literature, moreover, demonstrate that the increment of solution viscosity guarantees the formation of a uniform mat of nanofibers, without the presence of beads, known as one of the most common defects into the nanofibers mats [13]. It is widely explained how the molecular weight of polymer (Mw) and the polymeric concentration can control the presence of defects and diameter distributions inside the nanofiber mats. As the molecular weight increases, the number of beads and droplet is reduced. Since the increasing of the molecular weight can increase the instabilities distribution, the final nanofiber mats show a nonuniform distribution of diameters. Moreover, a low polymer concentration induces thinner fiber diameters, due to the evaporation of the solvent [14]. The direct correlation between polymeric concentration and viscosity modifies the jet deformations induced by viscoelastic forces during electrospinning. Therefore, when the polymeric concentration is too low, the electrospray process results to be the main electrified process deposition. On the contrary, when the solution viscosity is too high, during the electrospinning, the leak of charged jet from the tip of the needle could be compromised.

3.2. Electrospinning process parameters

All the parameters, related to the electrospinning process, such as voltage, flow rate, and working distance, tune the diameter distribution in the nanofiber mats, thus controlling the porosity distributions and the surface area of nanofibers.

Different works in the literature [3] demonstrated that the correlation between the voltage applied and the nanofiber morphology is not well defined. Nevertheless, this process parameter is quite important in order to establish, for each solution, the threshold value, above which the charged polymeric jet is originated, thus ensuring the nanofiber deposition on the counter electrode. Another fundamental parameter is the working distance, whose value can influence the completely evaporation of the solvent. Indeed, it is needed to define the minimum value of distance, able to provide the fiber's sufficient time to dry before depositing on the collector [15–18].

The flow rate is known as the rate at which the polymer solution is injected to the tip of the needle, defining then the flowing mass of solution and consequently the position of Taylor's cone related to the syringe needle. Moreover, a direct correlation between flow rate and the length of liner path, which precedes the bending instabilities, can be observed during the process [3]. At low values of flow rate, Taylor's cone is formed inside the tip of the needle, thus leading to an intermittent polymeric jet; however, at too high values of flow rates, the charged jet results to be continuous, leading to the formation of nanofiber mats, characterized by a large number of beads together with a nonuniform diameter distribution. Indeed, as the flow rate increases, the diameters of nanofibers mat increase [3].

This overview on all those parameters, which influence the morphology properties of nanofibers nets, provides important instruments in order to tune/control, during the electrospinning, the formation of some defects, i.e., nano-netting, able to optimize nanofiber mats, involved in several applications, like catalysis, sensors, optics, tissue engineering, and energy storage.

The nano-nets appear as spider-weblike structure, characterized by secondary ultrathin nanofibers interconnected with the main nanofibers [19]. There are some works in the literature [20–22] that study spider-weblike nanofibrous mat obtained by using an electrospun colloidal solution, containing the polymer and metal oxide nanoparticles. Kim et al. [20] obtained the nano-netting structure starting from a polymeric solution containing solid powder of ZnO mixed with a solution of nylon-6 and acetic acid. The ZnO nanoparticles induce a solution charge density increase, providing the separation of the thinner fibers from the main nanofiber web.

Moreover, Amna et al. [21] proposed the formation of secondary thinner web when ZnO nanoparticles were dissolved in a sol–gel solution of polyurethane in dimethylformamide (DMF).

4. Hollow nanostructures and coaxial electrospinning

One of the main important aspects of the electrospinning process is represented by the possibility to provide different types of nanofiber morphology, obtained by modifying the electrospinning technique. As an example, coaxial electrospinning is applied to the preparation of polymer core-shell nanofibers and hollow nanofibers composed not only of polymers but also of ceramics. Coaxial electrospinning is obtained by using two syringe supports disposed in a concentric configuration, and each syringe contains different spinning solutions, as sketched in **Figure 4(A)**. All parameters, described above, which influence the formation of polymeric jet during the electrospinning process, are the same. Coaxial electrospinning provides further advantages, when the molecular weight of polymer is too low to ensure the fiber formations, avoiding the droplets and consequently the electrospray process. Incorporating these kinds of polymer as the core into a core-shell nanofibers, it is possible to ensure the formation of a continuous jet and consequently the collection of nanofibers on the counter electrode. Moreover, core-shell fibers can offer a solution when it is needed to keep the functional components (proteins, enzymes, bacteria, viruses) maintaining their functionality. Core-shell nanofibers are characterized by a shell, based on solid materials, such as natural or synthetic polymers, and by a core, which is commonly a solvent (like water) with bio-systems. However, the hollow nanofibers are carried out when a wall is based on inorganic polymeric composites or ceramic materials, and the core results to be empty (as sketched in **Figure 4(B–D)**). There are two different approaches implemented to obtain hollow nanofibers. The first one is based on the concept of sacrificial polymer templates that is then removed. Choi et al. [23] fabricated hollow ZnO nanofibers by using polyvinyl alcohol (PVA) as polymeric template.

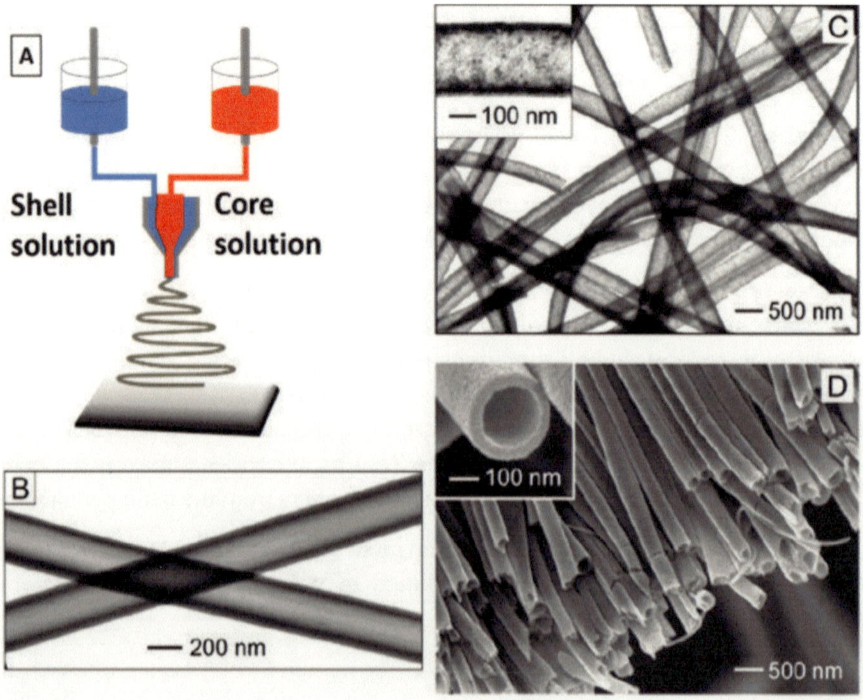

Figure 4. Core-shell nanofibers obtained by coaxial electrospinning (A) and hollow nanofibers (B-C-D) are proposed. The figure is adapted and reprinted with the permission from (Li and Xia [25]. Copyright (2004) American Chemical Society).

Another approach is based on coaxial electrospinning, starting from two immiscible liquids through the coaxial spinneret, followed by a selective removal of the core. Li et al. [24] studied hollow nanofibers, obtained by coaxial electrospinning, and used a polymeric solution of polyvinylpyrrolidone (PVP) and titania precursor (Ti $(OiPr)_4$) as shell and mineral oil as core. An example of the resulting hollow nanofiber is reported in **Figure 4(D)**.

Du et al. [22] used coaxial electrospinning in order to design TiO_2/ZnO core-shell nanofibers as photo-anodes in dye-sensitized solar cells (DSSCs). The resulting DSSC efficiency was close to 5%. This improvement can be related to the enhanced light-harvesting efficiency and electron collection efficiency.

5. Nanofiber deposition controlled by counter electrode and by patterning

The final step of electrospinning process is represented by the deposition of dried nanofiber mats on the counter electrode (collector). The collector is a conductive electrode, connected to the ground potential in order to provide a stable potential difference between the first electrode (tip of needle) and the second one (counter electrode). In electrospinning process, the deposition texture depends on the electrode configurations. Different works in the literature demonstrated the correlation between the morphological and physical properties of nanofibers with different types of counter electrodes. Indeed, different collectors can be divided into

(i) flat-plate collector, (ii) rotating drum collector, (iii) rotating wheel with edge, and (iv) parallel strips [4].

5.1. Flat-plate collector

In the majority of electrospinning setup, a flat-plate collector is used, thus leading to collect a non-woven nanofibers mats, namely as a random distribution of nanofibers on the counter electrode and on all substrate positioned on the top of it. The formation of a non-woven mat of nanofibers is induced by a layer-by-layer deposition on the planar surface. However, some applications required a certain alignment among the nanofibers. In order to induce certain fiber orientations, specific geometries of counter electrode, combined with its motion, are required (as proposed in **Figure 5**).

5.2. Rotating drum collector

Rotating cylindrical collectors combined with high rotating speed (up to 1000 rpm), as represented in **Figure 5(a)**, enhance a distribution of parallel nanofibers on it. In this configuration, the two components of velocity (rotating velocity and linear tangential velocity) play a key role in the alignment of nanofibers. The linear tangential velocity of each point on the collector surface is directly proportional to angular velocity and radius of cylinder. Therefore, when the tangential velocity assumes a threshold value able to guarantee the solvent evaporation of the jet, the nanofibers assumed a circular shape on the collector. However, if the tangential velocity results to be too low, random distribution of nanofibers occurred. Finally, if the tangential velocity is too high, the fiber jet will be broken, and the continuous nanofibers will no collect. Different works in the literature used drum collector in order to obtain aligned metal oxide nanofibers, based on TiO_2 [25, 26] and ZnO [23]. Aligned ZnO nanofibers are

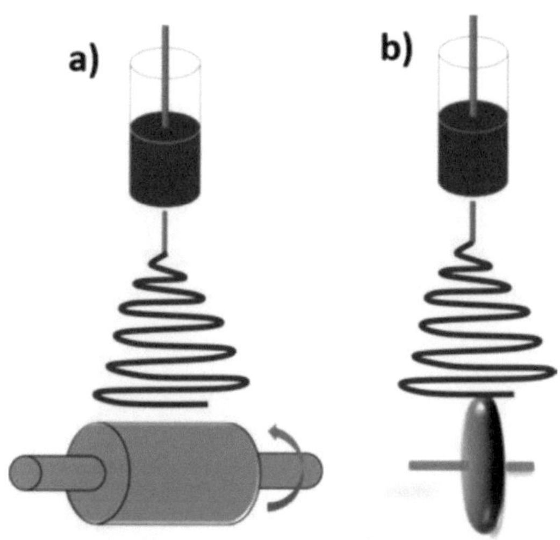

Figure 5. A schematic representation of different electrospinning collectors. In (a) and (b), the rotating drum and disk collectors are shown, allowing the aligned nanofiber mats.

electrospun starting from a gel containing the precursor of metal oxides and zinc acetate. The electrospun ordered nanofibers were then calcined at 450–500°C in oxygen atmosphere to induce the nucleation and growth of ZnO [23]. The aligned ceramic nanofibers show a higher surface-area-to-volume ratio, leading to enhance charge collection and their transport.

5.3. Rotating disk collector

The rotating disk can be defined as a thinner drum collector, on which the nanofibers are deposited on its edge. For this approach, collected nanofibers appear more aligned than the ones obtained with drum collector, described above [12, 27, 28]. **Figure 5(b)** represents a schematic view of rotating disk, and the nanofibers intercept the edge of counter electrode. With this kind of architecture of counter electrode, the rotation of the disk generates a tangential force, which acts on the polymeric jet, carrying out the nanofiber deposition only on the edge of the disk. This force reduces their diameter, stretching the nanofibers. In this configuration, the alignment of nanofibers results to be better than the one obtained by using rotating drum collector. However, the main limitation of this collector is that only a small quantity of aligned fibers can be obtained. In order to overcome the limitations induced by these kinds of counter electrode to obtain aligned nanofibers and to guarantee the formation of oriented nanofibers, several methods can be implemented. In particular, they mainly involve the modulation of external field provided by a specific geometry of counter electrode.

5.4. Patterning designed on counter electrode

Since some applications in energy field requires highly ordered structure, different works in the literature designed different patternings on a planar counter electrode, able to overcome all limitations introduced by different types of counter electrode (drum or rotating disk counter electrode) and enhance the aligned of nanofibers [29]. An example is shown in **Figure 6(A)**. Two gold bars are placed on the planar counter electrode, and their disposition breaks the asymmetry of the deposition, ensuring the deposition of parallel fibers. In a similar way, a quadripolar arrangement of isolated strips of electrodes, as represented in **Figure 6(B)**, induces a cross grating type of fiber deposition. Wu et al. [12] used two silver plate placed on the planar counter electrode, inducing a final aligned nanofiber mat with a highly ordered structure. The initial polymeric solution was made of CuO precursor mixed with PVA, dissolved in deionized water. The aligned CuO nanofibers, obtained after the calcination treatment conducted in air at a temperature of 500°C, enhanced their electrical transfer properties.

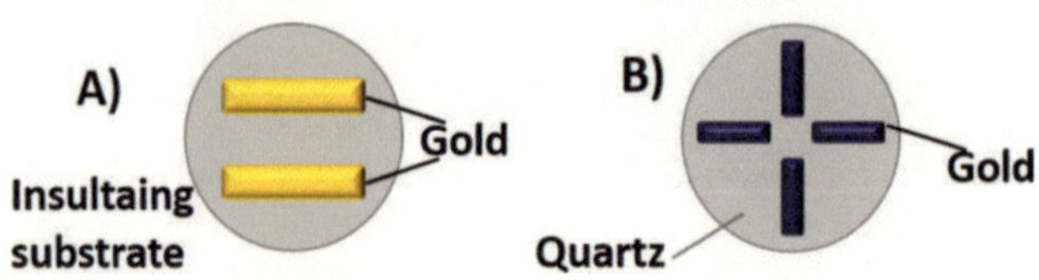

Figure 6. (A) Parallel arrangement of nanofibers induced by two gold strips, placed on planar counter electrode and (B) quadripolar arrangement of isolated strips of electrodes.

6. Application of semiconducting nanofibers

For the last decades, tremendous efforts have been devoted to the exploitation of renewable, efficient, and low environmental impact of energy sources, able to contribute to the rise of new models for a sustainable human development [30]. Nanostructured materials have demonstrated a huge potential in energy devices, significantly contributing to improve the final performance of the systems [31, 32]. Nanofibers by electrospinning belong to this intriguing class of materials. As described in the previous paragraphs, nanofibers offer a wide range of strategies to fine tune their morphology in order to meet the requirements of the final application. This versatility of the process provides the nanofibers with a huge potential for energy-related applications [33, 34].

Good examples exist, showing the integration of nanofibers in energy systems for both energy production and storage. In the following paragraphs, some examples are provided, especially focusing on photovoltaic systems and lithium-ion batteries.

6.1. Energy production

In photovoltaic devices, photons from solar light are directly converted into electrons thanks to the presence of proper materials. In traditional solar cells, adsorption/conversion is granted by a semiconducting material in a thin-film form, as GaAs, InP, and Si. In the most recently proposed systems, conversion is performed by organic molecules, as in dye-sensitized solar cells (DSSCs) [35, 36] or metalorganic lead halide perovskites in the so-called perovskite solar cells (PSCs) [37]. In both cases, a semiconductor is then needed to capture the generated electrons resulting in an electric current generation.

The leading wide bandgap, mesoporous semiconductor in these devices is TiO_2. Since the photo-generated charges have to be efficiently injected in the conduction band of the nano-structured semiconductor, the higher the injection efficiency, the lower the losses associated to the process. For this reason, the design of the semiconductor at the nanoscale plays a key role to obtain high-performing photo-electrodes. Nanofibers are quite promising candidates for obtaining well-performing devices since they offer several strategies to control and tune their morphology as needed by the final application.

As an example, the porosity of nanofiber mats can be considered: in DSSC, the possibility to control the dye uptake and the penetration of viscous, solid, or semisolid is a quite important feature. It is possible to change and control the nanofiber mat porosity, tuning the electrospinning process parameters, in order to optimize nanofiber mats for the design of photo-anodes in DSSC [38]. Nanofibers by electrospinning offer several strategies for low-temperature processing of photo-anodes: this is quite important for an easy integration of nanostructured semiconductors in plastic substrates for flexible devices [39, 40].

Several metal oxide semiconductors have been successfully fabricated by electrospinning, as TiO_2, ZnO, and SnO_2 [39, 40], for the design of well-performing photo-anodes.

Core-shell nanofibers [41]; TiO_2-graphene composite nanofibers [42]; electrospun ZnO photo-electrodes made of ZnO nanofibers with a dense, twisted structure [43]; and SnO_2/TiO_2 core-shell nanofiber-based photo-anodes have been successfully integrated in DSSC devices. Similarly, TiO_2- and Au-decorated TiO_2 nanofibers [44–46] have been proposed in PCSs.

Another important class of devices for energy conversion is represented by fuel cells. They are electrochemical devices, able to convert the chemical energy stored in several classes of molecules, acting as fuels (e.g., H_2, methanol, ethanol) into electricity in the presence of a catalyst. Different types of fuel cells exist according to the fuel that is converted (i.e., direct methanol fuel cells (DMFC)), the electrolyte that they use (i.e., solid oxide fuel cells (SOFCs)), or the catalyst that controls the oxidation process (i.e., microbial fuel cells (MFCs)).

In this area of energy, semiconducting nanofibers are especially proposed to design new cathode, when the oxygen reduction reaction (ORR) occurs at the cathode. As an example, manganese oxide nanofibers are successfully proposed to catalyze the ORR as an alternative to platinum [47]. One of the main disadvantages of metal oxides as catalysts to drive the ORR is related to their low electrical conductivity. To overcome this issue, different strategies have been proposed. A successful method is based on the use of composite nanofibers made of doped semiconductors, as proposed by Alvar et al. that optimized a process to embed carbon nanoparticles into mesoporous Nb-doped TiO_2 nanofibers [48].

6.2. Energy storage

In the area of energy storage, lithium-ion batteries (LIBs) play a crucial role as a promising technology toward sustainability. In a LIB, a negative electrode and a positive electrode are present, both able of reversibly intercalate Li+ions, and separated by a nonaqueous lithium-ion conducting electrolyte. During discharge, Li+ions carry the current from the negative to the positive electrode, through the nonaqueous electrolyte. During charge, an external high voltage is applied that forces lithium ions to migrate from the positive to the negative electrode, where the process known as intercalation occurs, during which they are embedded in the porous electrode material [31].

In this field, semiconducting nanofibers by electrospinning have been especially proposed for the fabrication of high-efficiency anodes. Good examples are represented by TiO_2 nanofibers. Han et al. fabricated TiO_2 hollow nanofibers sheathed with TiOxNy/TiN layers with the aim to optimize capability diffusion of lithium ions and electronic conductivity. The fabrication process was based on electrospinning to fabricate hollow nanofibers, followed by a thermal treatment in NH_3 atmosphere [49]. Another possibility is represented by the synthesis of composite TiO_2-based nanofibers. Zhang et al. proposed the fabrication by electrospinning, followed by a calcination step of TiO_2-graphene composite nanofibers able to behave as highly durable anodes [50].

Another interesting possibility offered by electrospinning is the decoration of carbon-based nanofibers with metal oxide catalysts, by adding the oxide precursor into the solution already containing the carbon precursor. The nucleation of the semiconducting oxide in the form of nanoparticles can then be achieved by the thermal process, which also permits the carbonization of the nanofibers. An interesting example of this process is offered by the work of Ji et al. [51]. They synthetized carbon nanofibers decorated with α-Fe_2O_3 nanoparticles, demonstrating homogenous dispersion of the nanoparticles along the carbon-based nanofibers. The composite mats were tested as anodes in Li-ion batteries; the resulting electrodes showed good reversibility and capacity.

High-capacity anodes for Li-ion batteries can be also designed using SnO_x, but strategies are needed for this semiconductor to improve its stability over cycles. Indeed, the variation of the volume induced by the intercalation process is detrimental for its mechanical stability, resulting in reduced lifetime of SnO_x-based anodes. In order to significantly improve the cycling durability of the resulting anodes, the electrospinning method is used to synthetize carbon-based nanofibers decorated with small-size SnO_x nanoparticles [52].

The possibility to use carbon-based nanofibers in the area of energy storage offers new interesting possibility to design flexible devices. Indeed, carbon-based mats can be processed to be freestanding and usually exhibit very high bendability, offering several possibilities of integration as electrodes in devices for smart electronics. In this area, several processes are developed to decorate the starting carbon mats with metal oxides to design new, well-performing anodes. Samuel et al. [53] decorated carbon-based nanofibers with MnO nanoparticles, demonstrating the possibility to couple the high performances achievable by this semiconducting oxide (923 mAh g^{-1} at a current rate of 123 mA g^{-1} after 90 cycles) with optimal flexibility of the carbon mats.

Author details

Giulia Massaglia[1,2]* and Marzia Quaglio[1]

*Address all correspondence to: giulia.massaglia@polito.it

1 Center For Sustainable Future Technologies@Polito, Istituto Italiano Di Tecnologia, Torino, Italy

2 Department of Applied Science and Technology (DISAT), Politecnico di Torino, Torino, Italy

References

[1] Jilani A, Abdel-Wahab MS, Hammad AH. Advance deposition techniques for thin film and. In: IntechOpen, editor. Modern Technologies for Creating the Thin-film Systems and coatings. Croatia: Intech; 2017. p. 137-149. DOI: 10.5772/65702

[2] Reneker DH, Yarin AL. Electrospinning jets and polymer nanofibers. Polymer. 2008;**49**: 2387-2425. DOI: 10.1016/j.polymer.2008.02.002

[3] Wendorff J, Agarwal S, Greiner A. Electrospinning: Materials, Processing and Applications. Weiheim Germany: Wiley-VCH; 201. 241 pp

[4] Huang ZM, Zhang YZ, Kotaki M, Ramakrishna S. A review on polymer nanofibers by electrospinning and their applications in nanocomposites. Composites Science and Technology. 2003;**63**:2223-2253. DOI: 10.1016/S0266-3538(03)00178-7

[5] Teo WE, Ramakrishna S. A review on electrospinning design and nanofibre assemblies. Nanotechnology. 2006;**17**:89-106. DOI: 10.1088/0957-4484/17/14/R01

[6] Bhardwa N, Kundu SC. Electrospinning: A fascinating fiber fabrication technique. Biotechnology Advances. 2010;**28**:325-347. DOI: 10.1016/j.biotechadv.2010.01.004

[7] Feng C, Khulbe KC, Matsuura T. Recent progress in the preparation, characterization, and applications of nanofibers and nanofiber membranes via electrospinning/interfacial polymerization. Journal of Applied Polymer Science. 2010;**115**:756-776. DOI: 10.1002/app.31059

[8] Hohman M, Ruletdge G, Brenner M. Electrospinning and electrically forced jets. II. Applications. Physics of Fluids. 2001;**13**:2221-2236. DOI: 10.1063/1.1384013

[9] Shin Y, Hohmann M, Brenner M, Ruletdge G. Experimental characterization of electrospinning: The electrically forced jet and instabilities. Polymer. 2001;**42**:9955-9967. DOI: 10.1016/S0032-3861(01)00540-7

[10] Reneker D, Yarin A, Fong H, Koombhongse S. Bending instability of electrically charged liquid jets of polymer solutions in electrospinning. Journal of Applied Physics. 2000;**87**:4531-4547. DOI: 10.1063/1.373532

[11] Doshi J, Reeneker D. Electrospinning process and applications of electrospun fibers. Journal of Electrostatics. 1995;**35**:151-160. DOI: 10.1016/0304-3886(95)00041-8

[12] Wu H, Li D, Pan W. Fabrication, assembly, and electrical characterization of CuO nanofibers. Applied Physics Letters. 2006;**89**:125-128. DOI: 10.1063/1.2355474

[13] Deitzel J, Kleinmeyer J, Harris D, Tan NCB. The effect of processing variables on the morphology of electrospun nanofibers and textiles. Polymer. 2001;**42**:261-272. DOI: 10.1016/S0032-3861(00)00250-0

[14] Bella F, Massaglia G, Chiodoni A, Pirri FC, Quaglio M. Dispelling clichés at the nanoscale: The true effect of polymer electrolytes on the performance of dye-sensitized solar cells. Nanoscale. 2015;**7**:12010-12017. DOI: 10.1039/C5NR02286J

[15] Megleski S, Stephens JS, Chase DB, Rabolt JF. Micro-and nanostructured surface morphology on electrospun polymer fibers. Macromolecules. 2002;**35**:8456-8466. DOI: 10.1021/ma020444a

[16] Frenot A, Chronakis IS. Polymer nanofibers assembled by electrospinning. Current Opinion in Colloid and Interface Science. 2003;**8**:64-75. DOI: 10.1016/S1359-0294(03)00004-9

[17] Beachley V, Wen X. Effect of electrospinning parameters on the nanofiber diameter and length. Materials Science and Engineering C. 2009;**29**:663-668. DOI: 10.1016/j.msec.2008.10.037

[18] Chronakis J. Novel nanocomposites and nanoceramics based on polymer nanofibers using electrospinning process—A review. Journal of Materials Processing Technology. 2005;**167**:283-293. DOI: 10.1016/j.jmatprotec.2005.06.053

[19] Massaglia G, Chiodoni A, Salvador GP, Delmondo L, Munoz-Tabares JA, Bocchini S, Sacco A, Bianco S, Saracco G, Quaglio M. Defining the role of nanonetting in the electrical

behaviour of composite nanofiber/nets. RSC Advances. 2017;**7**:38812-38818. DOI: 10.1039/C7RA05573K

[20] Kim HJ, Pant HR, Amarjargal A, Kim CS. Incorporation of silver-loaded ZnO rods into electrospun nylon-6 spider-web-like nanofibrous mat using hydrothermal process. Colloids and Surface A. 2013;**434**:49-55. DOI: 10.1016/j.colsurfa.2013.05.038

[21] Amna T, Hassan MS, Sheikh FA, Lee HK, Seo KS, Yoon D, Hwang IH. Zinc oxide- doped poly(urethane) spider web nanofibrous scaffold via one-step electrospinning: A novel matrix for tissue engineering. Applied Microbiology and Biotechnology. 2013;**97**:1725-1734. DOI: 10.1007/s00253-012-4353-0

[22] Du P, Song L, Xiong J, Li N, Xi Z, Wang L, Jin D, Guo S, Yuan Y. Coaxial electrospun TiO2/ZnO core–sheath nanofibers film: Novel structure for photoanode of dye-sensitized solar cells. Electrochimica Acta. 2012;**78**:392-397. DOI: 10.1016/j.electacta.2012.06.034

[23] Choi SH, Ankonina G, Youn DY, SG O, Hong JM, Kim ID. Hollow ZnO Nanofibers fabricated using electrospun polymer templates and their electronic transport properties. ACS Nano. 2009;**3**:2623-2631. DOI: 10.1021/nn900126k

[24] Li D, Xia Y. Direct fabrication of composite and ceramic hollow nanofibers by electrospinning. Nano Letters. 2004;**4**:933-938. DOI: 10.1021/nl049590f

[25] Shim HS, Na SI, Nam SH, Ahn HJ, Kim HJ, Kim DY, Kim WB. Efficient photovoltaic device fashioned of highly aligned multilayers of electrospun TiO2 nanowire array with conjugated polymer. Applied Physics Letters. 2008;**92**:183107-183110. DOI: 10.1063/1.2919800

[26] Tai QD, Chen B, Guo F, Xu S, Hu H, Sebo B, Zhao XZ. In situ prepared transparent polyaniline electrode and its application in bifacial dye-sensitized solar cells. ACS Nano. 2011;**20**:3795-3799. DOI: 10.1021/nn200133g

[27] Dersch R, Liu T, Schaper A, Greiner A, Wendorff J. Electrospun nanofibers: Internal structure and intrinsic orientation. Journal of polymer science. 2003;**41**:545-553. DOI: 10.1002/pola.10609

[28] Baji A, Mai Y, Wong S, Abtahi M, Chen P. Electrospinning of polymer nanofibers: Effects on oriented morphology, structures and tensile properties. Composites Science and Technology. 2010;**70**:703-718. DOI: 10.1016/j.compscitech.2010.01.010

[29] Fridrikh S, Yu J, Brenner M, Rutledge G. Controlling the fiber diameter during electrospinning. Physical Review Letter. 2001;**90**:144502-144504. DOI: 10.1103/PhysRevLett.90.144502

[30] Cao W, Hu Y. Utilisation and system integration. In: Renewable Energy, editor. IntechOpen; 2016

[31] Bianco S, Chiodoni A, Nair JR, Gerbaldi C, Quaglio M. Nanostructures for energy. In: Bhushan B, editor. Encyclopedia of Nanotechnology. Springer; 2016. pp. 2813-2827

[32] Aricò AS, Bruce P, Scrocati B, Tarascon JM, Van Schalkwijk W. Nanostructured materials for advanced energy conversion and storage devices. Nature Materials. 2005;**4**:366-377 DOI: 10.1038/nmat1368

[33] Kumar PS, Sundaramurthy J, Sundarrajan S, Babu VJ, Singh G, Allakhverdiev SI, Ramakrishna S. Hierarchical electrospun nanofibers for energy harvesting, production and environmental remediation. Energy and Environmental Science. 2014;**7**:3192-3222. DOI: 10.1039/C4EE00612G

[34] Massaglia G, Quaglio M. Semiconducting nanofibers in photoelectrochemistry. Materials Science in Semiconductor Processing. 2017;**73**:13-21. DOI: 10.1016/j.mssp.2017.06.047

[35] O'Regan B, Grätzel M. A low-cost, high-efficiency solar cell based on dye-sensitized colloidal TiO_2 films. Nature. 1991;**353**:737-740. DOI: 10.1038/353737a0

[36] Hardin BE, Snaith HJ, MD MG. The renaissance of dye-sensitized solar cells. Nature Photonics. 2012;**6**:162-169. DOI: 10.1038/nphoton.2012.22

[37] Sum TC, Mathews N. Advancements in perovskite solar cells: Photophysics behind the photovoltaics. Energy and Environmental Science. 2014;**7**:2518. DOI: 10.1039/C4EE00673A

[38] Dong Z, Kennedy SJ, Wu Y. Electrospinning materials for energy-related applications and devices. Journal of Power Sources. 2011;**196**:4886-4904. DOI: 10.1016/j.jpowsour.2011.01.090

[39] Joly D, Jung JW, Kim D, Demadrille R. Organophosphorus derivatives for electronic devices. Journal of Materials Chemistry C. 2016;**4**:10173. DOI: 10.1039/C6TC00590J

[40] Sundaramurthy J, Li N, Kumar PS, Ramakrishna S. Perspective of electrospun nanofibers in energy and environment. Biofuel Resources Journal. 2014;**2**:44-54. DOI: 10.18331/BRJ2015.1.2.3

[41] Miaoqiang L, Zheng D, Ye M, Guo W, Lai Y, Sun L, Lin C, Zuo J. Optimized porous rutile TiO2 nanorod arrays for enhancing the efficiency of dye-sensitized solar cells. Energy and Environmental Science. 2013;**6**:1615-1622. DOI: 10.1039/C3EE24125D

[42] He G, Wang X, Xi M, Zheng F, Zhu Z, Fong H. Fabrication and evaluation of dye-sensitized solar cells with photoanodes based on electrospun TiO2 nanotubes. Materials Letters. 2013;**106**:115-118. DOI: 10.1016/j.matlet.2013.05.014

[43] Madhavan AA, Kumar GG, Kalluri S, Joseph J, Nagarajan S, Nair S, Subramanian KRV, Balakrishnan KRV. Effect of embedded plasmonic au nanoparticles on photocatalysis of electrospun TiO2 nanofibers. Journal of Nanoscience and Nanotechnology. 2012;**12**:7963-7967

[44] Kim ID, Hong JM, Lee BH, Kim DY, Jeon EK, Choi DK, Yang DJ. Dye-sensitized solar cells using network structure of electrospun ZnO nanofiber mats. Applied Physics Letters. 2007;**91**:163109. DOI: 10.1063/1.2799581

[45] Dharani S, Mulmudi HK, Yantara N, Thu Trang PT, Park NG, Graetzel M, Mhaisalkar S, Mathews N, Boix PP. High efficiency electrospun TiO_2 nanofiber based hybrid organic–inorganic perovskite solar cell. Nanoscale. 2014;**6**:1675-1679. DOI: 10.1039/C3NR04857H

[46] Mali SS, Su Shim C, Kim H, Patil PS, Hong CK. *In situ* processed gold nanoparticle-embedded TiO2 nanofibers enabling plasmonic perovskite solar cells to exceed 14% conversion efficiency. Nanoscale. 2016;**8**:2664. DOI: 10.1039/C5NR07395B

[47] Delmondo L, Muñoz-Tabares JA, Sacco A, Garino N, Massaglia G, Castellino M, Salvador GP, Pirri CF, Quaglio M, Chiodoni A. Thermal evolution of Mn_xO_y nanofibres as catalysts for

the oxygen reduction reaction. Physical Chemistry Chemical Physics. 2017. DOI: 10.1039/C7C P05091G

[48] Alvar EN, Zhou B, Eichhorn SH. Carbon-embedded mesoporous Nb-doped TiO_2 nanofibers as catalyst support for the oxygen reduction reaction in PEM fuel cells. Journal of Materials Chemistry A. 2016;**4**:6540-6552. DOI: 10.1039/C5TA08801A

[49] Han H, Song T, Bae JY, Nazar LF, Kim H, Paik U. Nitridated TiO_2 hollow nanofibers as an anode material for high power lithium ion batteries. Energy and Environmental Science. 2011;**4**:4532-4536. DOI: 10.1039/C1EE02333K

[50] Zhang X, Kumar PS, Aravindan V, Liu HH, Sundaramurthy J, Mhaisalkar SG, Duong HM, Ramakrishna S, Madhavi S. Electrospun TiO_2−Graphene composite Nanofibers as a highly durable insertion anode for lithium ion batteries. The Journal of Physical Chemistry C. 2012;**116**:14780-14788. DOI: 10.1021/jp302574g

[51] Ji L, Toprakci O, Alcoutlabi M, Yao Y, Li Y, Zhang S, Guo B, Lin Z, Zhang X. α-Fe_2O_3 nanoparticle-loaded carbon nanofibers as stable and high-capacity anodes for rechargeable lithium-ion batteries. ACS Applied Materials and Interfaces. 2012;**4**:2672-2679. DOI: 10.1021/am300333s

[52] Zhu J, Lei D, Zhang D, Li G, Lu B, Wang T. Carbon and graphene double protection strategy to improve the SnO_x electrode performance anodes for lithium-ion batteries. Nanoscale. 2013;**5**:5499-5505. DOI: 10.1039/c3nr00467h

[53] Samuel E, Jo HS, Joshi B, An S, Park HG, Kim YI, Yoon WY, Yoon SS. Decoration of MnO Nanocrystals on flexible freestanding carbon Nanofibers for lithium ion battery anodes. Electrochimica Acta. 2017;**231**:582-589. DOI: 10.1016/j.electacta.2017.02.077